SpringerBriefs in Applied Sciences and Technology

Computational Mechanics

W0037023

Series Editors

Andreas Öchsner
Holm Altenbach
Lucas F. M. da Silva

For further volumes:
http://www.springer.com/series/8886

Eduardo N. Dvorkin · Rita G. Toscano

Finite Element Analysis of the Collapse and Post-Collapse Behavior of Steel Pipes: Applications to the Oil Industry

 Springer

Eduardo N. Dvorkin
Rita G. Toscano
SIM&TEC
Buenos Aires
Argentina

ISSN 2191-5342   ISSN 2191-5350 (electronic)
ISBN 978-3-642-37360-2   ISBN 978-3-642-37361-9 (eBook)
DOI 10.1007/978-3-642-37361-9
Springer Heidelberg New York Dordrecht London

Library of Congress Control Number: 2013934726

© The Author(s) 2013
This work is subject to copyright. All rights are reserved by the Publisher, whether the whole or part of
the material is concerned, specifically the rights of translation, reprinting, reuse of illustrations,
recitation, broadcasting, reproduction on microfilms or in any other physical way, and transmission or
information storage and retrieval, electronic adaptation, computer software, or by similar or dissimilar
methodology now known or hereafter developed. Exempted from this legal reservation are brief
excerpts in connection with reviews or scholarly analysis or material supplied specifically for the
purpose of being entered and executed on a computer system, for exclusive use by the purchaser of the
work. Duplication of this publication or parts thereof is permitted only under the provisions of
the Copyright Law of the Publisher's location, in its current version, and permission for use must
always be obtained from Springer. Permissions for use may be obtained through RightsLink at the
Copyright Clearance Center. Violations are liable to prosecution under the respective Copyright Law.
The use of general descriptive names, registered names, trademarks, service marks, etc. in this
publication does not imply, even in the absence of a specific statement, that such names are exempt
from the relevant protective laws and regulations and therefore free for general use.
While the advice and information in this book are believed to be true and accurate at the date of
publication, neither the authors nor the editors nor the publisher can accept any legal responsibility for
any errors or omissions that may be made. The publisher makes no warranty, express or implied, with
respect to the material contained herein.

Printed on acid-free paper

Springer is part of Springer Science+Business Media (www.springer.com)

To my daughters Cora and Julia
Eduardo N. Dvorkin

To my mother, Elsa
Rita G. Toscano

Contents

Chapter 1
Introduction

The production of oil and gas from offshore oil fields is, nowadays, more and more important. As a result of the increasing demand of oil, and being the shallow water reserves not enough, the industry is being pushed forward to develop and exploit more difficult fields in deeper waters.

As a consequence of the extremely severe work conditions, the constructors of deep-water pipelines need tubular products with enhanced resistance to withstand all the loads that will be applied to the pipeline, both during its construction and in operation. Among them: internal and external pressure, bending, tension, compression, concentrated loads, and thermal load combined with impact and fatigue.

A very important issue to take into account is the buckling phenomenon [1, 2]. The pipeline may buckle globally, either downwards (in a free span), horizontally ('snaking' on the seabed), or vertically (as upheaval buckling) (Fig. 1.1); global buckling implies buckling of the pipe as a bar in compression (column mode). If the internal pressure is higher than the external one, it introduces a destabilizing effect. It was demonstrated in [3, 4] that the pressurized pipe buckling load is lower than the Euler buckling load for the same pipe but under equilibrated internal/external pressures; on the other hand, when the external pressure is higher than the internal one, the resultant pressure has a stabilizing effect.

Steel pipes under external pressure may also reach their load carrying capacity due to a second failure mode: the localized collapse [5, 6]; in this case the pipe structure collapses with its sections losing their round shape (Fig. 1.2). The local buckling failure mechanism is most common during pipelay, due to excessive bending at the sagbend in conjunction with external overpressure. This book is focused on the collapse failure mode, considering pipes under external pressure only and external pressure plus bending.

In the design of marine pipelines it is very important to be able to determine the collapse pressure of steel pipes subjected to external hydrostatic pressure and bending. It is also required to be able to quantify the effect of shape imperfections, such as ovality and eccentricity, and of residual stresses on the collapse strength. The investigation of the post-collapse equilibrium path is also required to assess on the stability of this regime; that is to say, in order to assess if a collapse will be

E. N. Dvorkin and R. G. Toscano, *Finite Element Analysis of the Collapse and Post-Collapse Behavior of Steel Pipes: Applications to the Oil Industry*, SpringerBriefs in Computational Mechanics, DOI: 10.1007/978-3-642-37361-9_1,

Fig. 1.1 Global buckling

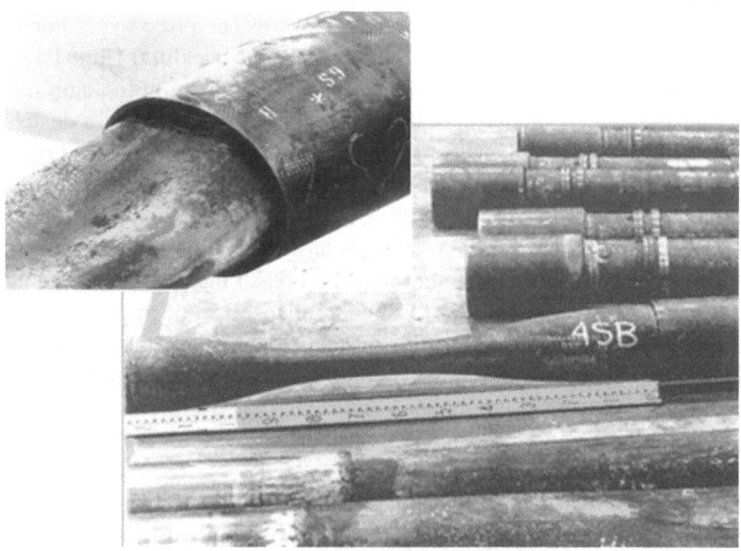

Fig. 1.2 Local buckling

localized in a section or will propagate along the pipeline. Therefore, it is also important to be able to quantify the effect of the geometrical imperfections and of the residual stresses on the post-collapse regime.

The external collapse pressure of very thin pipes is governed by the classical elastic buckling formulas [7, 8]; however for thicker pipes is necessary to take into account elasto-plastic considerations. The external collapse pressure depends on many factors, among them [9–14],

- the relation between the outside diameter and the wall thickness ($\frac{OD}{t}$ ratio),
- the yield stress of the pipe (σ_y) and its distribution through the pipe wall,
- the steel work-hardening characteristics,
- the pipe geometry (outside surface shape and wall thickness distribution along the pipe),
- the residual stresses (σ_R) locked in the pipe steel during the manufacturing process,
- the localized imperfections in the pipe body.

A pipeline that has been damaged locally or presents some localized imperfections, may locally collapse and, if the hydrostatic pressure is high enough, the collapse may propagate all along the pipeline. The collapse propagation pressure (P_p) is the lowest pressure value that can sustain the collapse propagation [5].

The external collapse propagation pressure is quite lower than the external collapse pressure; hence, in order to avoid the propagation of a collapse buckle and to limit the extent of damage to the pipeline, it is necessary to install buckle arrestors at intervals along the pipeline [15]. Buckle arrestors are devices that locally increase the bending stiffness of the pipe. There are many different types of arrestors but all of them typically take the form of thick-walled rings. The external pressure necessary for propagating the collapse pressure through the buckle arrestors is the collapse cross-over pressure (the minimum pressure value at which the buckle crosses over the arrestor).

In Fig. 1.3 we show a drawing of the arrestor we are studying: the integral ring buckle arrestor; together with its characteristic curve [*Externalpressurevs.Internalvol.variation*], where we can identify the collapse pressure, the collapse propagation pressure and the cross-over pressure. We also see the photographs of two arrestors after collapse; in the first case, obviously the external pressure was lower than the cross-over pressure, as the collapse did not cross over the arrestor and the downstream pipe is not collapsed. In the second case, on the contrary, the external pressure was higher than the cross-over pressure, and both, upstream and downstream pipes are collapsed.

Two different integral buckle arrestor cross-over mechanisms were identified in the literature: flattening and flipping. The occurrence of either cross-over mechanism is determined by the geometry of the pipes and of the arrestors [16, 17]. Experimental results and numerical analyses are available in the literature for the cross-over of integral ring buckle arrestors under external pressure, on different diameters and materials [15, 16, 18–21]. To validate our numerical results on buckling arresting and cross-over mechanisms, we performed a series of laboratory tests on medium-size carbon steel pipes.

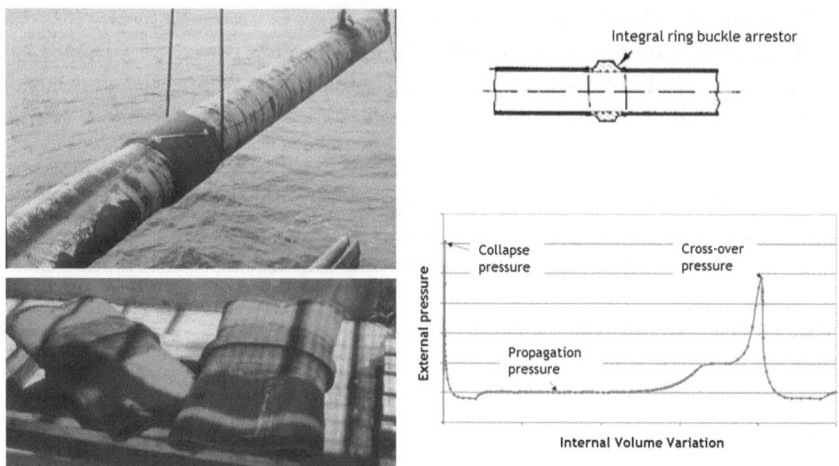

Fig. 1.3 Integral ring buckle arrestor

From a computational mechanics viewpoint, the topic of tracking the collapse and post-collapse equilibrium path of elasto-plastic structures such as steel pipes, which are very sensitive to imperfections and that can develop finite strains during its deformation, is very challenging. Even though much research has been dedicated in the last twenty years to this subject, there is still ample room for improvement.

Since for the purposes described above numerical models are a fundamental engineering tool, it is of utmost importance to have reliable models that can accurately simulate the pre- and post-collapse behavior of steel pipes under external pressure and bending. In order to assure the reliability of the developed numerical models, it is essential to validate them, by comparing the finite element results with experimental ones.

In order to model the behavior of steel pipes under external pressure and bending, and numerically describe the phenomena of collapse and post-collapse buckles propagation, we use nonlinear shell finite elements. In Chap. 2 we discuss shell element formulations for general nonlinear analysis.

In Chap. 3 we discuss the determination of the external collapse pressure of steel pipes using finite element models and the effect of some imperfections on the structural behavior of the pipes. We describe some simple 2D finite element models developed as a first approach for the simulation of the external pressure collapse test. We compare the results provided by these 2D models with the experimental results; the comparison demonstrates that a 2D geometrical characterization of the pipes does not contain enough information to accurately assess on their collapse strength. However, the 2D models are a useful tool for performing parametric studies on the effects of ovality, eccentricity, residual stresses and work-hardening on the external collapse pressure of the pipes.

To include bending in our analysis, we developed a 3D model of infinite pipes, using MITC4 shell elements that include shear deformation [22–24]. This model accurately describes the behavior of very long pipes. Considering the pipe dimensions under analysis (in the cases that we analyze radius/thickness >8), the use of shell elements to model the pipes seems to be appropriate, as the validations that we performed confirm. We have already presented the results and conclusions shown in this chapter in Refs. [6, 25, 26].

But the pipe shape imperfections and the wall thickness normally change along its length; also localized imperfections can be found in the pipes. Therefore, we also implemented 3D finite element models of finite pipes, developed to overcome the limitations of the simpler models described previously. The geometrical information on the test samples was acquired using the imperfections measuring system (IMS or "shapemeter"), and were used as data for our 3D finite element models. The description of the IMS is included in Appendix A [6, 27–29].

Several tests programs were developed, where we compared the finite element results with the experimental ones obtained at different labs, such as C-FER (Edmonton, Canada) and Tenaris Siderca (Campana, Argentina). In Chap. 4 we discuss the experimental program that we developed to validate our numerical models for predicting the collapse and post-collapse behavior of steel pipes under external pressure and bending [26, 30, 31]. We also discuss the application of our numerical models to the analysis of the collapse of slotted pipes and of steel pipes under external pressure and axial load [32].

In Chap. 5 we present the analysis of the collapse and post-collapse behavior of pipelines reinforced with buckle arrestors. We developed finite element models to analyze the collapse, collapse propagation and cross-over mechanisms of reinforced pipes under external pressure only and we present an experimental validation of the models. In particular we studied the case of welded integral arrestors [33, 34]. The comparisons between the numerical and experimental results show that the developed finite element models are able to simulate the flattening and the flipping cross-over mechanisms.

References

1. Palmer AC, King RA (2004) Subsea pipeline engineering. PennWell Corporation, Tulsa
2. Kyriakides S, Corona E (2007) Mechanics of offshore pipelines, vol I. Buckling and collapse. Elsevier
3. Palmer AC, Baldry JAS (1974) Lateral buckling of axially constyrained pipelines. J Petroleum Tech 26:1283-1284
4. Dvorkin EN, Toscano RG (2001) Effects of external/internal pressure on the global buckling of pipelines. In: Bathe K-J (ed) Computational fluid and solid mechanics—proceedings first MIT conference on computational fluid and solid mechanics. Elsevier
5. Palmer AC, Martin JH (1975) Buckle propagation in submarine pipelines. Nature 254:46–48
6. Assanelli AP, Toscano RG, Johnson D, Dvorkin EN (2000) Experimental/numerical analysis of the collpase behavior of steel pipes. Eng Comput 17:459–486
7. Timoshenko SP, Gere JM (1961) Theory of elastic stability. McGraw Hill, New York

8. Brush DO, Almroth BO (1975) Buckling of bars, plates and shells. McGraw Hill, New York
9. Heise O, Esztergar EP (1970) Elasto-plastic collapse under external pressure. ASME J Eng Ind 92:735–742
10. Clinedinst WO (1977) Analysis of collapse test data and development of new collapse resistance formulas. Report to the API Task Group on performance properties
11. Fowler JR, Klementich EF, Chappell JF (1983) Analysis and testing of factors affecting collapse performance of casing. ASME J Energy Resour Technol 105:574–579
12. Kanda M, Yazaki Y, Yamamoto K, Higashiyama H, Sato T, Inoue T, Murata T, Morikawa H, Yanagimoto S (1983) Development of NT-series oil-country tubular good. Nippon Steel Techn Rep 21:247–262
13. Krug G (1983) Testing of casing under extreme loads. Insitute of Petroleum Engineering Technische Universtät, Clausthal
14. Mimura H, Tamano T, Mimaki T (1987) Finite element analysis of collapse strength of casing. Nippon Steel Technical Rep 34:62–69
15. Johns TG, Mesloh RE, Sorenson JE (1978) Propagating buckle arrestors for offshore pipelines. ASME J Press Vessel Technol 100:206–214
16. Park TD, Kyriakides S (1997) On the performance of integral buckle arrestors for offshore pipelines. Int J Mech Sci 39:643–669
17. Kyriakides S, Park TD, Netto TA (1998) On the design of integral buckle arrestors for offshore pipelines. Int J Appl Ocean Res 20:95–104
18. Netto TA, Estefen SF (1996) Buckle arrestors for deepwater pipelines. Int J Mar Struct 9:873–883
19. Langer CG (1999) Buckle arrestors for deepwater pipelines. In: Proceedings of the offshore technology conference, OTC 10711, Houston, Tx
20. Netto TA, Kyriakides S (2000) Dynamic performance of integral buckle arrestors for offshore pipelines. Part I: Experiments. Int J Mech Sci 42:1405–1423
21. Netto TA, Kyriakides S (2000) Dynamic performance of integral buckle arrestors for offshore pipelines. Part II: Analysis. Int J Mech Sci 42:1425–1452
22. Dvorkin EN, Bathe K-J (1984) A continuum mechanics based four-node shell element for general nonlinear analysis. Eng Comput 1:77–88
23. Bathe K-J, Dvorkin EN (1985) A four-node plate bending element based on Mindlin/Reissner plate theory and a mixed interpolation. Int J Numer Methods Eng 21:367–383
24. Bathe K-J, Dvorkin EN (1986) A formulation of general shell elements—the use of mixed interpolation of tensorial components. Int J Numer Methods Eng 22:697–722
25. Toscano RG, Amenta PM, Dvorkin EN (2002) Enhancement of the collapse resistance of tubular products for deepwater pipeline applications. In: IBC'S offshore pipeline Technology, conference documentation
26. Toscano RG, Timms C, Dvorkin EN, DeGeer D (2003) Determination of the collapse and propagation pressure of ultra-deepwater pipelines. In: OMAE 2003, 22nd international conference on offshore mechanics and artic engineering
27. Arbocz J, Babcock CD (1969) The effect of general imperfections on the buckling of cylindrical shell. ASME J Appl Mech 36:28–38
28. Arbocz J, Williams JG (1977) Imperfection surveys of a 10-ft diameter shell atructure. AIAA J 15:949–956
29. Yeh MK, Kyriakides S (1988) Collapse of deepwater pipelines. ASME J Energy Res Tech 110:1–11
30. Toscano RG, Gonzalez M, Dvorkin EN (2003) Validation of a finite element model that simulates the behavior of steel pipes under external pressure. J Pipeline Integr 2:74–84
31. Toscano RG, Mantovano LO, Dvorkin EN (2004) On the numerical calculation of collapse propagation pressure of steel deepwater pipelines under external pressure and bending: experimental verification of the finite element results. In: Proceedings 4th international conference on pipeline technology, pp 1417–1428
32. Toscano RG, Dvorkin EN (2011) Collapse of steel pipes under external pressure and axial tension. J Pipeline Eng 4:213–214

33. Toscano RG, Mantovano LO, Amenta P, Charreau R, Johnson D, Assanelli AP, Dvorkin EN (2006) Collapse arrestors for deepwater pipelines: finite element models and experimental validation for different cross-over mechanisms. In: Proceedings OMAE 2006—25th International conference on offshorw mechanics and artic engineering. Hamburg, Germany
34. Toscano RG, Mantovano LO, Amenta P, Charreau R, Johnson D, Assanelli AP, Dvorkin EN (2008) Collapse arrestors for deepwater pipelines. Cross-over mechanisms. Comput Struct 86:728–743

Chapter 2
Shell Element Formulations for General Nonlinear Analysis. Modeling Techniques

2.1 Introduction

In 1970, Ahmad, Irons and Zienkiewicz [1] presented a shell element formulation that after many years still constitutes the basis for modern finite element analysis of shell structures. The original formulation was afterwards extended to material and geometric nonlinear analysis under the constraint of the infinitesimal strains assumption [2–4].

The fundamental features of the A-I-Z shell element are,

- using isoparametric interpolation functions the displacements inside the shell element are interpolated from three displacement-d.o.f. and two rotation-d.o.f. at each node,
- the interpolated generalized displacement fields present C^0 continuity,
- the element is not based on any plate/shell theory but it is a continuum element incorporating several assumptions that we list below (degenerated solid element).

The kinematic and constitutive assumptions are,

- a straight line that is initially normal to the mid-surface remains straight after the deformation,
- a straight line that is initially normal to the mid-surface is not stretched during the deformation,
- the through-the-thickness stresses are zero.

It is important to remark that the second assumption precludes the consideration of finite strain kinematics.

Although the A-I-Z shell element was a breakthrough in the field of finite element analysis of shell structures, it suffers from the locking phenomenon and much research effort has been devoted to the development of A-I-Z type elements that do not incorporate this problem [5, 6].

The MITC4 shell element [7–9] which was developed to overcome the locking problem of the A-I-Z shell elements has become, since its development in the early

E. N. Dvorkin and R. G. Toscano, *Finite Element Analysis of the Collapse and Post-Collapse Behavior of Steel Pipes: Applications to the Oil Industry*, SpringerBriefs in Computational Mechanics, DOI: 10.1007/978-3-642-37361-9_2, © The Author(s) 2013

1980s, the standard shell element for many finite element codes. However, the limitation of infinitesimal strains is still present in the MITC4 formulation.

Many researchers have contributed to the development of shell elements that can model finite strain situations, among them,

- an early contribution by Rodal and Witmer for elasto-viscoplastic material models (J_2) where, at each iteration, after going through the displacements calculation, the shell element thickness is updated neglecting the elastic strains and invoking the incompressibility of the viscoplastic flow (J_2) [10],
- in 1983 Hughes and Carnoy [11] developed a finite strain shell element for the Mooney-Rivlin material model which uses a plane-stress constitutive relation for the laminae and updates afterwards the thickness via a staggered iterative formulation,
- Simo and co-workers in the period 1988–1992 developed a complete 3D nonlinear shell element formulation [12–16],
- Ramm and co-workers developed 3D shell elements considering also through-the-thickness stretching [17, 18].

In 1995 Dvorkin et al. developed the MITC4-TLH element, that based on the original MITC4 formulation can model finite strain elasto-plastic (J_2) deformations. This element imposes the condition of zero transversal stresses and its computational cost was rather high [19, 20].

Later, Toscano and Dvorkin developed an element that is also based on the MITC4 formulation and can efficiently model finite strain deformations using a general 3D material model: the MITC4-3D element [21, 22].

The most relevant differences with the original MITC4 formulation are:

- for each quadrilateral element there are 22 d.o.f.: 5 generalized displacements per node plus 2 extra d.o.f. to incorporate the through-the-thickness stretching, these extra d.o.f. are condensed at the element level;
- a general 3D constitutive relation is used, instead of the original laminae plane stress constitutive relation.

2.2 The Standard A-I-Z Quadrilateral Shell Element for Linear Analysis

2.2.1 Linear Analysis Kinematics

When modeling a shell we define, on its mid-surface, *nodes* and at those nodes we define *director vectors* which are the best approximation to the shell mid-surface normal at the corresponding nodes. The A-I-Z quadrilateral element is defined using four nodes which are not necessarily coplanar.

Fig. 2.1 Kinematics of the
A-I-Z shell element

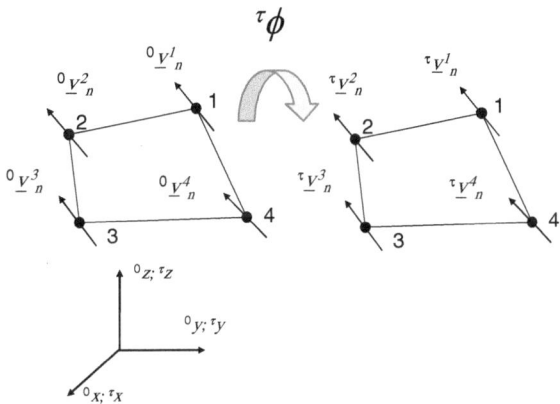

Under the assumption of infinitesimal strains, for a configuration at a time τ, a point inside the shell element, with natural coordinates [5] (r, s, t) (see Fig. 2.1), is defined by the Cartesian coordinates,[1]

$$^{\tau}\underline{x}(r,\, s,\, t) = h_k(r, s)\; ^{\tau}\underline{x}_k + \frac{t}{2}h_k(r, s)[a^{\tau}\underline{V}^n]_k. \tag{2.1}$$

In the above equation,

h_k 2D isoparametric interpolation function corresponding to the k-node [5];

$^{\tau}\underline{x}_k$ position vector of the mid-surface k-node at time τ;

$a|_k$ shell thickness at the k-node (assumed as invariant during the deformation);

$^{\tau}\underline{V}^n|_k$ director vector at the k-node at time τ $\left(\left\|^{\tau}\underline{V}_n|_k\right\| = 1\right)$;

while the natural coordinates (r, s) are defined on the element mid-surface $(t = 0)$ the natural coordinate t is measured at any point along the corresponding director vector direction. The second term on the r.h.s. in Eq. (2.1) shows that at any point on the element mid-surface the unit director vector times the thickness is interpolated from the nodal values.

The geometry interpolation in Eq. (2.1) presents C^0 continuity.

For describing the kinematics of the A-I-Z element the two main assumptions are:

- the element thickness remains constant due to the assumed infinitesimal strains deformation;
- the director vectors remain straight during the deformation.

The covariant base vectors of the (r, s, t) system are determined deriving Eq. (2.1),

---

[1] We use Einstein's notation: $a_k b_k \equiv \sum_k a_k b_k$, that is to say repeated indices indicate a summation.

$$\tau_{\underline{g}_i} = \frac{\partial\,^\tau\underline{x}}{\partial r_i} \tag{2.2}$$

and the contravariant base vectors need to fulfill the relation,

$$\tau_{\underline{g}}^i \cdot \tau_{\underline{g}_j} = \delta_j^i \tag{2.3}$$

For linear kinematics we consider the τ-configuration to be coincident with the 0-(reference) configuration.

The displacements for a time τ are,

$$^\tau\underline{u} = {}^\tau\underline{x} - {}^0\underline{x}. \tag{2.4}$$

Using the above kinematic assumptions we get,

$$^\tau\underline{u} = h_k\,^\tau\underline{u}_k + \frac{t}{2}h_k\left[a\left(^\tau\underline{V}_n - {}^0\underline{V}_n\right)\right]_k. \tag{2.5}$$

Let us now define in the 0-configuration at the node k two vectors that with the nodal director vector form the ortho-normal basis $(^0\underline{V}_1, {}^0\underline{V}_2, {}^0\underline{V}_n)$. We can write, for infinitesimal rotations [23, 24],

$$\begin{aligned}
^\tau\underline{V}_n^k &= {}^0\underline{V}_n^k + {}^\tau\underline{\theta}_k \times {}^0\underline{V}_n^k \\
^\tau\underline{\theta}_k &= \alpha_k\,{}^0\underline{V}_1^k + \beta_k\,{}^0\underline{V}_2^k \\
^\tau\underline{V}_n^k &= {}^0\underline{V}_n^k + \beta_k\,{}^0\underline{V}_1^k - \alpha_k\,{}^0\underline{V}_2^k.
\end{aligned} \tag{2.6}$$

Therefore,

$$^\tau\underline{u} = h_k\,^\tau\underline{u}_k + \frac{t}{2}h_k\left[a\left(-\alpha\,{}^0\underline{V}_2 + \beta\,{}^0\underline{V}_1\right)\right]_k. \tag{2.7}$$

It is apparent from Eq. (2.7) that this element formulation introduces 5 d.o.f. per node.

At any point inside the shell we can write the infinitesimal strain tensor in terms of its covariant components $(\tilde{\varepsilon}_{lm})$ in the (r, s, t) curvilinear system and the corresponding contravariant base vectors,

$$\underline{\varepsilon} = \tilde{\varepsilon}_{ij}\,{}^0\underline{g}^i\,{}^0\underline{g}^j \tag{2.8}$$

where we use the notation $(^0\underline{g}^i\,{}^0\underline{g}^j)$, to indicate the dyadic (tensorial) product between the two contravariant base vectors.[2] In Eq. (2.8), $\tilde{\varepsilon}_{tt} = 0$ because the thickness is constant.

From the kinematic relations between strain components and displacements [24] we get,

---

[2] Some authors use the notation $^0\underline{g}^i \otimes {}^0\underline{g}^j$.

$$\tilde{\varepsilon}_{ij} = \frac{1}{2}\left({}^0\underline{g}_i \cdot \frac{\partial \underline{u}}{\partial r_i} + {}^0\underline{g}_j \cdot \frac{\partial \underline{u}}{\partial r_i}\right).$$ (2.9)

Hence, using Voigt notation we can write

$$\tilde{\underline{\varepsilon}} = \tilde{\underline{B}}\,\underline{U}$$ (2.10)

where $\tilde{\underline{\varepsilon}}$ is the (5×1) column vector formed with the non-zero curvilinear components of the strain tensor, $\underline{U}$ is the (20×1) column vector with the element nodal generalized displacements and $\tilde{\underline{B}}$ is the (5×20) strain-displacement matrix, formed using Eq. (2.9) [7].

2.2.2 Stress-Strain Relations

The assumption of zero stresses through the thickness is equivalent to consider that each surface parallel to the mid-surface is in a plane stress condition. In the A-I-Z finite element discretization, with only C^0 continuity, there are two alternative ways for imposing through the shell thickness the plane stress condition,

- imposing it to the different laminae with constant t;
- imposing it at every point to the surfaces normal to the director vector.

We have chosen the second alternative.

At each point inside the shell element we define the local Cartesian system $(\hat{\underline{e}}_1, \hat{\underline{e}}_2, \hat{\underline{e}}_3)$ with,

$$\hat{\underline{e}}_1 = \frac{{}^0\underline{g}_2 \times {}^0\underline{g}_3}{\left\|{}^0\underline{g}_2 \times {}^0\underline{g}_3\right\|}$$

$$\hat{\underline{e}}_3 = \frac{{}^0\underline{g}_3}{\left\|{}^0\underline{g}_3\right\|}$$ (2.11)

$$\hat{\underline{e}}_2 = \hat{\underline{e}}_3 \times \hat{\underline{e}}_1.$$

In this local Cartesian system we formulate the different plane stress constitutive relations in the plane $(\hat{\underline{e}}_1, \hat{\underline{e}}_2)$.

There is an obvious contradiction since the above defined plane stress state, due to the imposed kinematic constraint, is also a plane strain state; this can only be possible in a very specific orthotropic material model. We overlook this contradiction as the price that we pay for degenerating the solid into a shell element.

The constitutive tensor can be described as,

$$\underline{\underline{C}} = C_{ijkl}\,\hat{\underline{e}}_i\,\hat{\underline{e}}_j\,\hat{\underline{e}}_k\,\hat{\underline{e}}_l = \tilde{C}^{pqrs}\,{}^0\underline{g}_p\,{}^0\underline{g}_q\,{}^0\underline{g}_r\,{}^0\underline{g}_s$$ (2.12)

and from the above we can get the curvilinear components $\widetilde{C}^{pqrs}$.

Then the element stiffness matrix can be calculated as [7],

$$\underline{K} = \int_V \underline{\widetilde{B}}^T \, \underline{\widetilde{C}} \, \underline{\widetilde{B}} \, dV. \tag{2.13}$$

The (5×5) matrix $\underline{\widetilde{C}}$ collects the curvilinear components of the constitutive tensor and it is symmetric for hyperelastic materials models and elasto-plastic material models when an associated plasticity model is used [24].

2.2.3 The Locking Problem

The locking problem has been very much analyzed in the literature [5, 6]; in the present section we just present a couple of very simple examples to illustrate it.

2.2.3.1 Shear Locking

Using a 4-node element to model the cantilever under constant moment in Fig. 2.2 we notice that,

- the u_2 displacement interpolation is linear along the coordinate x_1 with a zero at node 1,
- the θ rotation interpolation is linear with a zero at node 1.

The shear deformation $\gamma = \frac{du_2}{dx_1} - \theta$ has to be zero everywhere and the condition is imposed more strongly when the thickness tends to zero [20, 25].

It is evident that considering the order of the interpolation functions and the boundary condition, the only solution is $u_2 = const = 0$.

Fig. 2.2 Cantilever under constant moment modeled with one 4-node element: shear locking

2.2.3.2 Shear and Membrane Locking

When using a parabolic interpolation for the generalized displacements of a curved cantilever we find the impossibility of imposing together the zero shear and zero membrane elongation conditions.

2.2.4 Solving the Locking Problem

The first remedy that was proposed for the locking problem was the use of reduced or selective integration schemes [6]; however, those schemes, even though they are very simple and produce inexpensive elements, incorporate the difficulty of the spurious rigid body modes and the oscillation in the stress predictions [5, 6].

The element MITC4 was developed by Dvorkin and Bathe as a solution for the shear locking problem that does not incorporate numerical drawbacks.

2.3 The MITC4 Quadrilateral Shell Element for Linear Analysis

This element incorporates the displacement/rotation interpolations used in the A-I-Z element; the curvilinear covariant strain components $(\widetilde{\varepsilon}_{rr}, \widetilde{\varepsilon}_{ss}, \widetilde{\varepsilon}_{rs})$ are directly calculated from the displacement/rotation interpolations using Eq. (2.9).

For the out-of-surface shear components we use the interpolations in Fig. 2.3, which can be written as,

$$
\begin{aligned}
\widetilde{\varepsilon}_{rt} &= \frac{1}{2}\left(1+s\right)\widetilde{\varepsilon}_{rt}\big|_A^{DI} + \frac{1}{2}\left(1-s\right)\widetilde{\varepsilon}_{rt}\big|_C^{DI}, \\
\widetilde{\varepsilon}_{st} &= \frac{1}{2}\left(1+r\right)\widetilde{\varepsilon}_{st}\big|_D^{DI} + \frac{1}{2}\left(1-r\right)\widetilde{\varepsilon}_{st}\big|_B^{DI}.
\end{aligned}
\tag{2.14}
$$

In Eq. (2.14) we use the notation,

$\widetilde{\varepsilon}_{ij}\big|_P^{DI}$ covariant strain component calculated from the displacement interpolations at the *sampling point* P

The element, defined as we describe in this section, satisfies the Patch Test, does not present spurious rigid body modes and does not lock. In the literature there is abundant numerical evidence on the element robustness and accuracy.

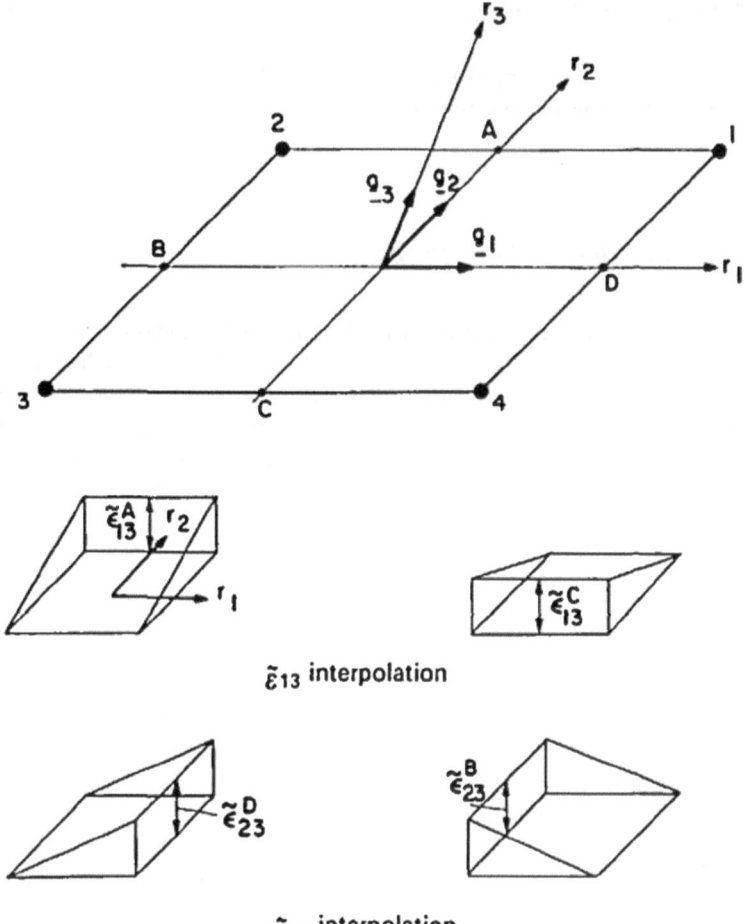

$\tilde{\varepsilon}_{13}$ interpolation

$\tilde{\varepsilon}_{23}$ interpolation

Fig. 2.3 The MITC4 out-of surface shear interpolations

2.4 Nonlinear Analysis Using the MITC4 Element

2.4.1 Infinitesimal Strains Problems: Total Lagrangean Formulation

There are a number of nonlinear structural and mechanical problems for which the infinitesimal strains approach provides acceptable results. For these cases we developed for the MITC4 element a Total Lagrangean Formulation (TLF) [5].

In an incremental analysis we know the τ-configuration and we seek the $(\tau + \Delta\tau)$-configuration. Using the Principle of Virtual Work we can state for the equilibrium at the configuration at $(\tau + \Delta\tau)$ [24],

$$\int_{^0V} {}^{\tau+\Delta\tau}_0\underline{\underline{S}} : \delta^{\tau+\Delta\tau}_0\underline{\underline{\varepsilon}} \, {}^0dV = {}^{\tau+\Delta\tau}\delta W^{ext}. \tag{2.15}$$

In the above equation $^0\underline{V}$ is the volume of the reference configuration (*time* $= 0$); and for the configuration at $(\tau + \Delta\tau)$ ${}^{\tau+\Delta\tau}_0\underline{\underline{S}}$ is the second Piola-Kirchhoff stress tensor; ${}^{\tau+\Delta\tau}_0\underline{\underline{\varepsilon}}$ is the Green-Lagrange strain tensor (both tensors referred to the reference configuration) and ${}^{\tau+\Delta\tau}\delta W^{ext}$ is the virtual work of the external loads.

We can write for the incremental step,

$$\begin{aligned}{}^{\tau+\Delta\tau}_0\underline{\underline{S}} &= {}^{\tau}_0\underline{\underline{S}} + {}_0\underline{\underline{S}}, \\ {}^{\tau+\Delta\tau}_0\underline{\underline{\varepsilon}} &= {}^{\tau}\varepsilon + {}_0\underline{\underline{\varepsilon}}.\end{aligned} \tag{2.16}$$

Where the tensors ${}_0\underline{\underline{S}}$; ${}_0\underline{\underline{\varepsilon}}$ are increments referred to (*time* $= 0$).

We decompose the strain increment into a linear (${}_0\underline{\underline{e}}$) and nonlinear part (${}_0\eta$) in terms of the incremental displacements and relate the incremental strain tensor with the incremental stress tensor using and incremental fourth order constitutive tensor ${}_0\underline{\underline{\underline{C}}}$.

After linearizing we get for the incremental step [24],

$$\int_{^0V} \delta_0\underline{\underline{e}} : {}_0\underline{\underline{\underline{C}}} : {}_0\underline{\underline{e}} \, {}^0dV + \int_{^0V} {}^{\tau}_0\underline{\underline{S}} : \delta_0\eta^0 dV = {}^{\tau+\Delta\tau}\delta W^{ext} - \int_{^0V} {}^{\tau}_0\underline{\underline{S}} : \delta_0\underline{\underline{e}} \, {}^0dV. \tag{2.17}$$

As it is well known, the $(\tau + \Delta\tau)$-configuration is determined iterating on Eq. (2.15) until equilibrium is fulfilled [5].

In the nonlinear problem, we need to handle finite rotations; hence, we can write [23],

$$^{\tau+\Delta\tau}\underline{V}^k_n = {}^{\tau+\Delta\tau}_0\underline{\underline{R}} \cdot {}^0\underline{V}^k_n. \tag{2.18}$$

Any rotation matrix can be written as [23],

$$^{\tau}_0\underline{\underline{R}} = \underline{\underline{I}}_3 + \frac{\sin\theta_k}{\theta_k}\underline{\underline{\Theta}}_k + \frac{1}{2}\left(\frac{\sin\left(\theta_k/2\right)}{\left(\theta_k/2\right)}\right)^2 \underline{\underline{\Theta}}^2_k \tag{2.19}$$

where,

$$\theta_k = \left(\alpha^2_k + \beta^2_k\right)^{1/2}$$

$$[\Theta_k] = \begin{bmatrix} 0 & 0 & \beta_k \\ 0 & 0 & -\alpha_k \\ -\beta_k & \alpha_k & 0 \end{bmatrix}. \tag{2.20}$$

As in the infinitesimal rotations case, we have 5 d.o.f./node.

In the case of finite rotations the linearized equilibrium equations present extra terms that were discussed in Refs. [20, 23].

2.4.2 Finite Strains

For the analysis of finite strain problems we use the following interpolation for the reference configuration geometry [21, 22],

$$
{}^{0}\underline{x}(r, s, t) = h_k(r, s)\, {}^{0}\underline{x}_k + \frac{t}{2}\, {}^{0}\underline{d}a \tag{2.21}
$$

where,

$$
{}^{0}\underline{d} = \frac{h_k(r,\, s)\, {}^{0}\underline{V}^n|_k}{\left\| h_k(r,\, s)\, {}^{0}\underline{V}^n|_k \right\|}. \tag{2.22}
$$

It is important to remark that we considered in Eq. (2.21) elements with uniform thickness.

For the displacement field we use,

$$
{}^{\tau}u(r,\, s,\, t) = h_k(r,\, s)\, {}^{\tau}\underline{u}_k + \frac{t}{2}\left({}^{\tau}\lambda_o + {}^{\tau}\lambda_1 t \right)\left({}^{\tau}\underline{d} - {}^{0}\underline{d} \right)a \tag{2.23}
$$

where,

$$
{}^{\tau}\underline{d} = \frac{h_k(r,\, s)^{\tau}\underline{V}^n|_k}{\left\| h_k(r,\, s)^{\tau}\underline{V}^n|_k \right\|}. \tag{2.24}
$$

Equations (2.22) and (2.24) are used to avoid spurious director vector stretching [26] that in this finite strains case, in which we do not neglect the stretching through the thickness, may affect the results.

In Eq. (2.23) ${}^{\tau}\lambda_o$ is the constant thickness stretching and ${}^{\tau}\lambda_1$ is the through-the-thickness stretching gradient. In our formulation both stretching d.o.f. are condensed at the element level.

The strain interpolations are the same as the ones we used in the infinitesimal strains case. However, for hyperelastic material models we interpolate the Green-Lagrange covariant strain tensor components and for the elasto-plastic material we interpolate the covariant components of the Hencky (logarithmic) strain tensor [24].

We use 3D constitutive relations; hence, the through-the-thickness stress component is not neglected. In [22] we developed the 3D constitutive equations for the elasto-plastic element based on Lee's multiplicative decomposition of the deformation gradient tensor and maximum energy dissipation [24, 27].

2.5 Modeling Considerations

In this section we discuss several considerations that need to be taken into account when modeling shell structures.

2.5.1 The Nodal Director Vectors

The nodal director vectors may be either defined by the analyst or calculated by the finite element code.

When the analyst introduces the director vectors together with the mid-surface nodes, she/he selects them at each node so as to be the best approximation to the actual normal to the shell mid-surface.

When the finite element code calculates at a given node the director vectors, the normals to the interpolated mid-surface are calculated for all the elements sharing the node; hence, the code defines at the node as many director vectors as the number of elements sharing the node (the interpolated mid-surfaces only have C^0 continuity). All the element normals sharing a node rotate together.

2.5.2 Number of d.o.f. per Node

As discussed above, in a shell model there may be nodes at which only one director vector is defined and nodes at which multiple director vectors are defined.

For the case of nodes with only one director vector, the analyst has to consider 5 d.o.f. (3 displacements and 2 rotations around the local axes $^\tau\underline{V}_1$ and $^\tau\underline{V}_2$).

For the case of nodes with multiple director vectors, the analyst has to consider 6 d.o.f. (3 displacements and 3 rotations around the global Cartesian axes).

The case of a node with multiple but very close director vectors has to be treated as a case with 5 d.o.f. collapsing the very close director vectors.

It is important to be aware of the fact that when using 6 d.o.f. the rotational boundary conditions need to be defined along the global Cartesian axes and when using 5 d.o.f. the rotational boundary conditions need to be defined along the local axes. In geometrically nonlinear analyses these local axes change for each incremental step.

When modeling a stiffened shell using shell elements and iso-beam (Timoshenko beam) elements [5],[3] at the nodes shared by a shell and a beam element 6 d.o.f. have to be used.

---

[3] Please notice that shell elements are not compatible with Bernoulli beam elements.

References

1. Ahmad S, Irons B, Zienkiewicz O (1970) Analysis of thick and thin shell structures by curved finite elements. Int J Numer Methods Eng 2:419–451
2. Ramm E (1977) A plate/shell element for large deflections and rotations. In: Bathe et al (ed) Formulations and computational algorithms in finite element analysis. MIT Press, Cambridge
3. Kråkeland B (1978) Nonlinear analysis of shells using degenerate isoparametric elements. In: Bergan et al (ed), Finite elements in nonlinear mechanics. Tapir Publishers, Norwegian Institute of Technology, Trondheim
4. Bathe K-J, Bolourchi S (1980) A geometric and material nonlinear plate and shell element. Comput Struct 11:23–48
5. Bathe K-J (1996) Finite element procedures. Prentice Hall, Saddle River
6. Zienkiewicz O, Taylor R (2000) The finite element method. Butterworth-Heinemann, Oxford
7. Dvorkin EN, Bathe K-J (1984) A continuum mechanics based four-node shell element for general nonlinear analysis. Eng Comput 1:77–88
8. Bathe K-J, Dvorkin EN (1985) A four-node plate bending element based on Mindlin/Reissner plate theory and a mixed interpolation. Int J Numer Methods Eng 21:367–383
9. Bathe K-J, Dvorkin EN (1986) A formulation of general shell elements—the use of mixed interpolation of tensorial components. Int J Numer Methods Eng 22:697–722
10. Rodal J, Witmer E (1979) Finite-strain large-deflection elastic-viscoplastic finite-element transient analysis of structure. NASA CR 159874
11. Hughes T, Carnoy E (1983) Nonlinear finite element shell formulation accounting for large membrane strains. Comput Methods Appl Mech Eng 39:69–82
12. Simo J, Fox D (1989) On a stress resultant geometrically exact shell model. Part I: Formulation and optimal parametrization. Comput Methods Appl Mech Eng 72:267–304
13. Simo J, Fox D, Rifai M (1989) On a stress resultant geometrically exact shell model. Part II: The linear theory; computational aspects. Comput Methods Appl Mech Eng 72:53–92
14. Simo J, Fox D, Rifai M (1990) On a stress resultant geometrically exact shell model. Part III: Computational aspects of the nonlinear theory. Comput Methods Appl Mechs Eng 79:21–70
15. Simo J, Fox D, Rifai M (1992) On a stress resultant geometrically exact shell model. Part IV: Variable thickness shells with through-the-thickness stretching. Comput Methods Appl Mech Eng 81:91–126
16. Simo J, Kennedy J (1992) On a stress resultant geometrically exact shell model. Part V: Nonlinear plasticity formulation and integration algorithms. Comput Methods Appl Mech Eng 96:133–171
17. Büchter M, Ramm E, Roehl D (1994) Three-dimensional extension of non-linear shell formulation based on the enhanced assumed strain concept. Int J Numer Methods Eng 37:2551–2568
18. Bischoff M, Ramm E (1997) Shear deformable shell elements for large strains and rotations. Int J Numer Methods Eng 40:4427–4449
19. Dvorkin EN, Pantuso D, Repetto E (1995) A formulation of the MITC4 shell element for finite strain elasto-plastic analysis. Comput Methods Appl Mech Eng 125:17–40
20. Dvorkin EN (1995) Nonlinear analysis of shells using the MITC formulation. Arch Comput Methods En 2:1–50
21. Toscano RG, Dvorkin EN (2007) A shell element for finite strain analyses. Hyperelastic material models. Eng Comput 24:514–535
22. Toscano RG, Dvorkin EN (2008) A new shell element for elasto-plastic finite strain analyzes. Application to the collapse and post-collapse analysis of marine pipelines. In: Abel J, Cooke J (eds), Proceedings 6th international conference on computation of shell & spatial structures, Spanning Nano to Mega. Ithaca
23. Dvorkin EN, Oñate E, Oliver X (1988) On a nonlinear formulation for curved Timoshenko beam elements considering large displacement/rotation increments. Int J Numer Methods Eng 26:1597–1613

24. Dvorkin EN, Goldschmit MB (2005) Nonlinear continua. Springer, Berlin
25. Dvorkin EN (1992) On nonlinear analysis of shells using finite elements based on mixed interpolation of tensorial components. In: Rammerstorfer F (ed) Nonlinear analysis of shells by finite elements. Springer, New York
26. Gebhardt H, Schweizerhof K (1993) Interpolation of curved shell geometries by low order finite elements—errors and modifications. Int J Numer Methods Eng 36:287–302
27. Simo J, Hughes T (1998) Computational inelasticity. Springer, New York

Chapter 3
Collapse and Post-Collapse Behavior of Steel Pipes. Finite Element Models

3.1 Introduction

The main objective of this chapter is to discuss some basic ideas regarding the behavior of steel pipes under external pressure and bending.

In Sect. 3.2 we describe some simple 2D finite element models that we developed as a first approach for the simulation of the external pressure collapse test. We compare the results provided by these 2D models with the experimental results obtained at an experimental lab. The comparison demonstrates that a 2D geometrical characterization of the pipes does not contain enough information to assess on their collapse strength. However, the 2D models are a useful tool for performing parametric studies on the effects of the ovality, eccentricity, residual stresses and work-hardening of the pipes on the external collapse pressure values.

In Sect. 3.3 we develop a 3D model to describe the behavior of very long pipes. This model can simulate the behavior of the pipes not only under external pressure but also under bending.

In Sect. 3.4 we present the 3D finite element models of finite pipes, and we use it to study the measurement of the residual stresses. In the next chapter we will incorporate in the model a proper description of the pipe geometry.

3.2 Two Dimensional Finite Element Models of Very Long Pipes

In this section we discuss the 2D finite element models that we implemented to simulate the behavior of ideal long specimens in the external pressure collapse test. We also compare experimental results with the predictions of these 2D models.

E. N. Dvorkin and R. G. Toscano, *Finite Element Analysis of the Collapse and Post-Collapse Behavior of Steel Pipes: Applications to the Oil Industry*, SpringerBriefs in Computational Mechanics, DOI: 10.1007/978-3-642-37361-9_3, © The Author(s) 2013

3.2.1 Formulation of the 2D Models

We develop the 2D finite element models using a total Lagrangian formulation [1] that incorporates,

- geometrical nonlinearity due to large displacements/rotations (infinitesimal strains assumption),
- material nonlinearity, an elasto-plastic constitutive relation is used for modeling the steel mechanical behavior (von Mises associated plasticity [2]).

We develop the finite element analyses using a special version of the general purpose finite element code ADINA [3] that incorporates the quadrilateral QMITC element [4–6].

For modeling the external hydrostatic pressure we use follower loads [7], and we introduce in our models the residual stresses with a linear distribution across the thickness.

When using a 2D model it is important to recognize that the actual collapse test is not modeled exactly neither by plane strain nor by plane stress models because the absence of longitudinal restraints imposes a plane stress situation at the sample edges and the length L of the samples (L/D >10) approximates a plane strain situation at its center.

In this section, in order to explore the limitations of the 2D models we analyze the collapse test using both, plane stress and plane strain finite element models.

3.2.2 Two Dimensional Finite Element Results Versus Experimental Results

The 2D finite element models are developed considering an elastic-perfectly plastic material constitutive relation. We will show in Sect. 3.2.3 that disregarding the steel work-hardening introduces only a negligible error in the calculated collapse pressures.

For standard tests, the laboratory keeps, for each sample, records of the average, maximum and minimum outside diameter (D) at three sections (the central section and the two end sections) and thicknesses (t) at the two end sections (eight points per section).

We construct the geometry of the 2D finite element models using,

- the ovality (Ov) and the average outside diameter of the central section,
- the eccentricity (ε) obtained by averaging the eccentricities of the two end sections and the average thickness obtained by averaging the sixteen thickness determinations.

The ovality and eccentricity are defined as,

$$Ov = \frac{D_{\max} - D_{\min}}{D_{average}}$$

$$\varepsilon = \frac{t_{\max} - t_{\min}}{t_{average}}.$$

The residual stresses for each sample are measured using a slit-ring test. The actual transversal yield stress of the sample material is measured and its value is used for the elastic-perfectly plastic material constitutive relation.

To determine the collapse loads of the sample models we calculate the non-linear load-displacement path and seek for its horizontal tangent.

To perform the numerical analyses we use a 2D mesh with 720 QMITC elements (Fig. 3.1) and 1572 d.o.f. Half of the pipe is modeled due to symmetry. To assess on the quality of this mesh we analyze the plane strain collapse of an infinite pipe and we compare our numerical results with the analytical results obtained using the formulas in Ref. [8].

From the results in Table 3.1, we conclude that the proposed 2D mesh of QMITC elements is accurate enough to represent the collapse of very long specimens.

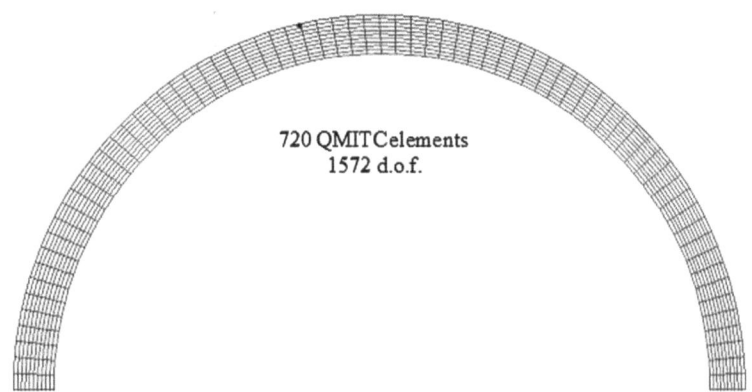

720 QMITC elements
1572 d.o.f.

Fig. 3.1 2D finite element mesh

Table 3.1 Qualification of the 2D finite element model	Average OD (mm)	245.42
	Average thickness [mm]	12.61
	$\frac{D}{t}$	19.47
	Ov (%)	0.18
	σ_y (Mpa)	890
	$\frac{\text{Theoretical\_result}}{\text{FE\_result}}$	0.992

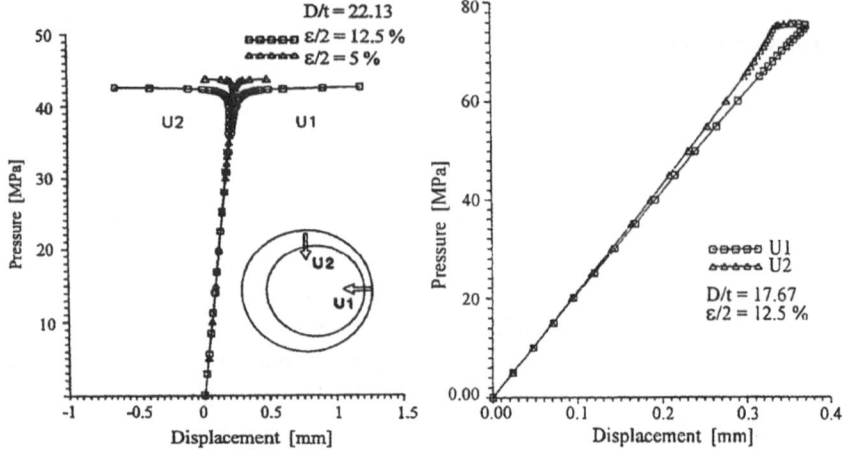

Fig. 3.2 Eccentric tube: direct collapse behavior (D/t = 22.13) and inverse collapse behavior (D/t = 17.67)

From the analyses of different cases we identified two basic types of load/ displacement paths. In Fig. 3.2 we present both, direct and inverse collapse behavior of eccentric pipes. The latter case was previously identified in the literature for the case of collapse under external pressure and bending [9, 10].

A total of 32 collapse tests, for casings 9 5/8″ OD 47 lb/ft and 7″ OD 26 lb/ft both Grade 95 ksi, were analyzed using plane stress and plane strain models. The comparisons between the numerical and experimental results are plotted in Fig. 3.3.

It is important to observe that the collapse pressure values determined using the 2D models present a significant deviation from the experimental ones, being in general the former lower than the latter. Some reasons for this behavior are,

- the middle section ovality is not fully representative of the sample geometry,
- in developing the 2D models the measured ovality is entirely assigned to the shape of the first elastic buckling mode. This conservative approach partially accounts for the fact that the numerical values are lower than the actual ones,

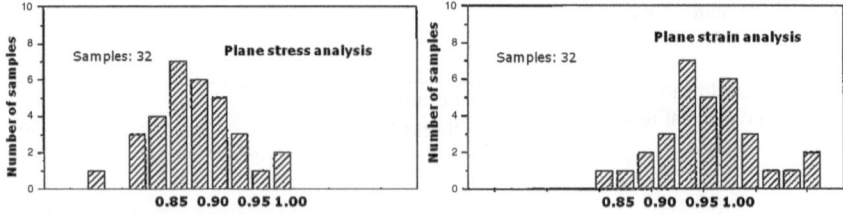

Fig. 3.3 Qualification of the 2D models. FEA collapse pressure/experimental collapse pressure. 9 5/8″ OD 47# and 7″ OD 26# SD95HC

- the experimental set-up used in these tests (Fig. 3.4) imposes on the samples unilateral radial restraints at both ends. These restraints are not described by the 2D models; the difference between the numerical and actual boundary conditions also partially accounts for the fact that the collapse pressures predicted by the 2D models are in general lower than the actual ones.

It is obvious, from the results in Fig. 3.3, that the 2D geometrical characterization of a pipe is not enough for assessing on its collapse performance. However, the 2D models are very useful for performing parametric studies on the relative weight of different factors that affect the external collapse pressure.

3.2.3 Strain Hardening Effect

It was already mentioned that the strain hardening of the casing material does not play an important role in the determination of its external collapse pressure. In this section we examine the above assessment using 2D finite element models.

We analyze two 9 5/8″ pipes (a thin and a thick one) using a bilinear material model. In order to explore different hardening values we consider three values for the constant plastic tangential modulus: $E_t = 0.0$ (perfect plasticity), 0.057E and 0.10E, where E is the Young modulus. We also consider two ovality (Ov) values but we do not include in this analysis neither the pipe eccentricity nor its residual stresses. We summarize the model predicted collapse pressures in the cells of Table 3.2.

As it can be seen, for all the analyzed cases, the strain hardening has a negligible effect on the external collapse pressure value.

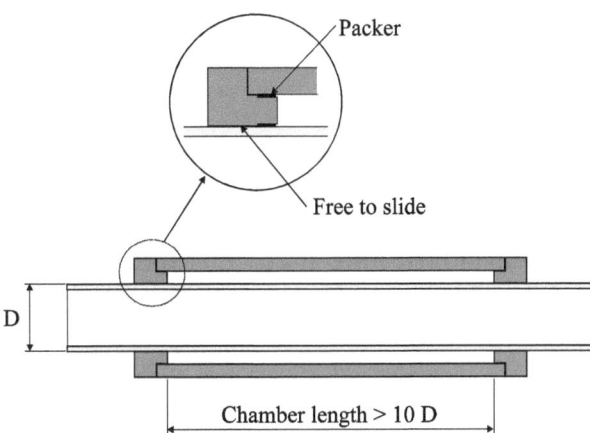

Fig. 3.4 Collapse chamber

Table 3.2 Analysis of the work-hardening effect on the external collapse pressure (MPa)

$\frac{D}{t}$	Ov (%)	$E_t = 0.0$	$E_t = 0.057E$	$E_t = 0.1E$
17.66	0.75	59.9	60.2	60.4
17.66	0.35	67.1	67.3	67.5
24.37	0.75	26.7	26.7	26.7
24.37	0.35	29.0	29.1	29.1

It is important to recognize that in the above analyses we considered σ_y to be independent of E_t. This is not the case if as yield stress we adopt the one corresponding to a relatively large permanent offset [11].

3.2.4 Effect of Ovality, Eccentricity and Residual Stresses

Using 2D finite element models we conducted a parametric study aimed at the analysis of the effect, on the casing collapse pressure, of the ovality (Ov), eccentricity (ε) and residual stresses (σ_R).

In the present analyses the ovality is considered to be concentrated in the shape corresponding to the first elastic buckling mode and the eccentricity is modeled considering non-coincident OD and ID centers.

In Fig. 3.5 we plot the results of our parametric study, normalized with the collapse pressure calculated according to API Bulletin 5C3 (1994). It is obvious from these results that the main influence on the external collapse pressure comes from the ovality and from the residual stresses; however, the effect of the residual stresses diminishes when the ratio (D/t) evolves from the plastic collapse range to the elastic collapse range. The eccentricity effect, in the case of the external collapse test with neither axial nor bending loads, is minor.

Most of the literature dealing with casing collapse pressure agrees on the importance of the ovality effect [10–14].

As a conclusion of the above parametric study we can assess that for producing pipes with enhanced collapse pressure it is very important to have both, a low ovality and low residual stresses. In actual production processes these two requirements are usually conflictive.

3.3 Three Dimensional Finite Element Model of Very Long Pipes

In the previous section we presented finite element models developed for studying the collapse behavior of long steel pipes under external pressure; our purpose in the present section is to extend the study to the post-collapse regime and to loading cases that combine external pressure and bending [15].

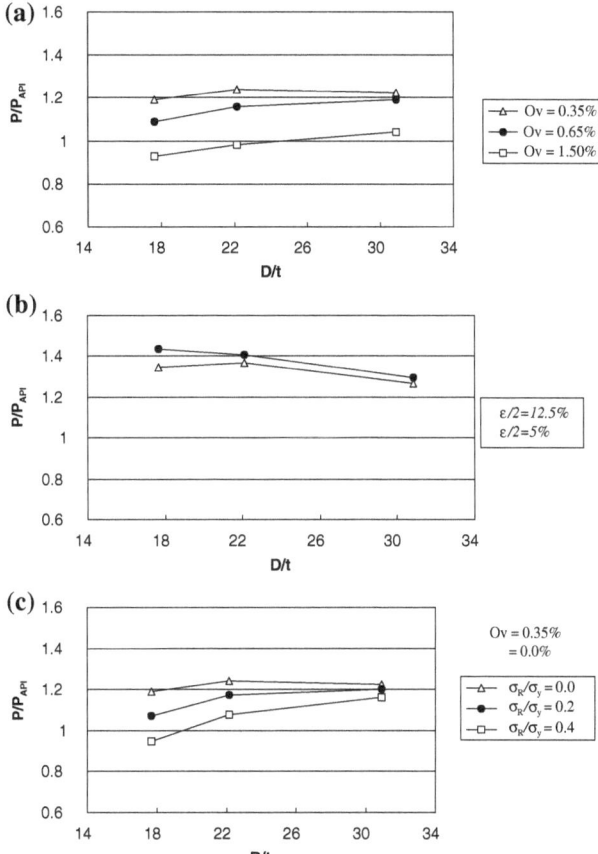

Fig. 3.5 Parametric study of the effect of ovality, eccentricity and residual stresses on the casing collapse pressure. **a** Ovality effect. **b** Eccentricity effect. **c** Residual stressess effect

We develop a numerical model to simulate the behavior of a very long pipe (infinite pipe) and determine its pre and post-collapse equilibrium path. Using this model we perform parametric studies in order to investigate the significance of the different geometrical imperfections and of the residual stresses on the collapse and collapse propagation pressures.

3.3.1 Formulation of a 3D Model for Very Long Pipes

The finite element models were developed using the nonlinear shell elements in the general-purpose finite element code ADINA [3]. The main features of the finite element models are,

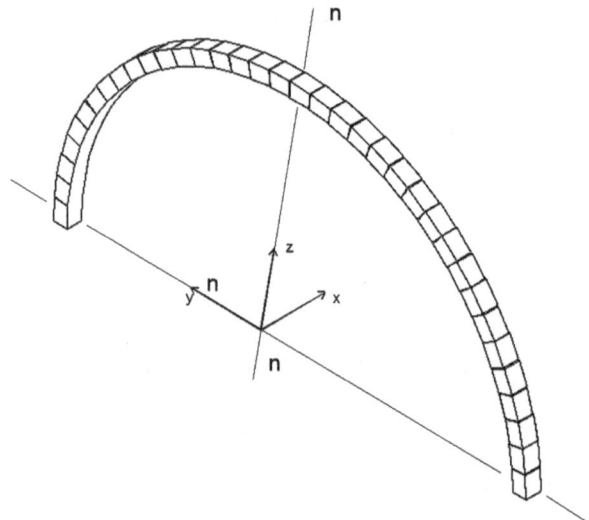

Fig. 3.6 Finite element mesh for the analysis of very long pipes

- MITC4 shell element (4-node element that includes shear deformations) [16–18],
- automatic solution of the incremental nonlinear finite element equations [19],
- material non-linearity: elasto-plastic material with multilinear hardening, associated plasticity according with the von Mises yield rule and isotropic or kinematic hardening [1],
- geometrical nonlinearity: large displacements/rotations [1].

For the cases with external pressure plus bending we first impose the bending and then the external pressure keeping constant the imposed curvature.

In Fig. 3.6 we present the finite element mesh, which we use for the analysis of very long pipes (the evolution of the cross section shape does not depend on the axial coordinate).

3.3.1.1 Boundary Conditions

- In one of the transverse planes ($x = constant$), both the axial displacements and the rotations respect to n–n *axis* are restricted.
- The other transverse plane is free to move in the *x-direction*, but is restricted to stay flat (see the constraint equations below).
- Rotations are restricted according to the symmetry conditions with respect to the plane $z = 0$.
- In order to avoid rigid body displacements, a minimum number of additional constraints are applied.

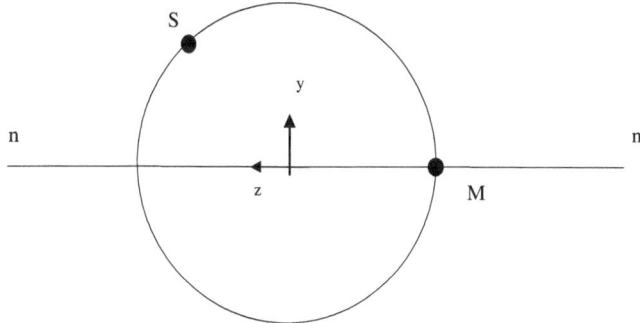

Fig. 3.7 Master and slave nodes on the axially unrestricted section

To be sure that the section keeps plane when bending is applied, we impose constraint equations.

Figure 3.7 shows the "Master" and "Slave" nodes for those nodes on the section with unrestricted axial displacements.

The master node M is one of the two nodes on the intersection between the cross section and the neutral axis corresponding to the applied bending, on the axially unrestricted section. All the other nodes in that section are the slave nodes S.

The constraint equations are,

$$u_S = u_M + \theta_Z^M (y_S - y_M) + \theta_Z^M (v_S - v_M)$$

$$\theta_Z^S = \theta_Z^M$$

In the above equations,
y Initial y coordinate,
u Displacement along the x axis,
v Displacement along the y axis,
θ_z Rotation with respect to the n–n axis.

It is important to take into account that when the sample is long enough (L/D >10) the end conditions have only a very small influence on the collapse pressure [20].

Following with the example described in Fig. 3.6, we compare the results obtained using the two different shell models,

- Short model with no axial displacements (shell under plane strain conditions)
- (L/D = 10) model with the ends restrained to remain on a plane (welded end cups).

The results summarized in Table 3.3 indicate the equivalence of both models. For cases with (L/D <10) we may expect the end conditions to play a more significant role.

Table 3.3 Long shell model compared with plane strain shell model

Short model under plane strain conditions	$\dfrac{\text{Theretical\_result}}{\text{FE\_shell\_PS}} = 0.980$
Long model: $\dfrac{L}{D} = 10$	$\dfrac{\text{Theretical\_result}}{\text{FE\_shell\_long}} = 0.978$

3.3.2 Validation of the Finite Element Model

In order to validate the finite element model, in the present section we analyze a series of cases presented by Kyriakides in [21].

Even though the pipes analyzed by Kyriakides are of smaller outer diameter and different material characteristics (aluminum) than the pipes that we analyze with our model, the mechanism of the collapse behavior is identical; hence, after the validation of the model, we use it on typical steel seamless pipes.

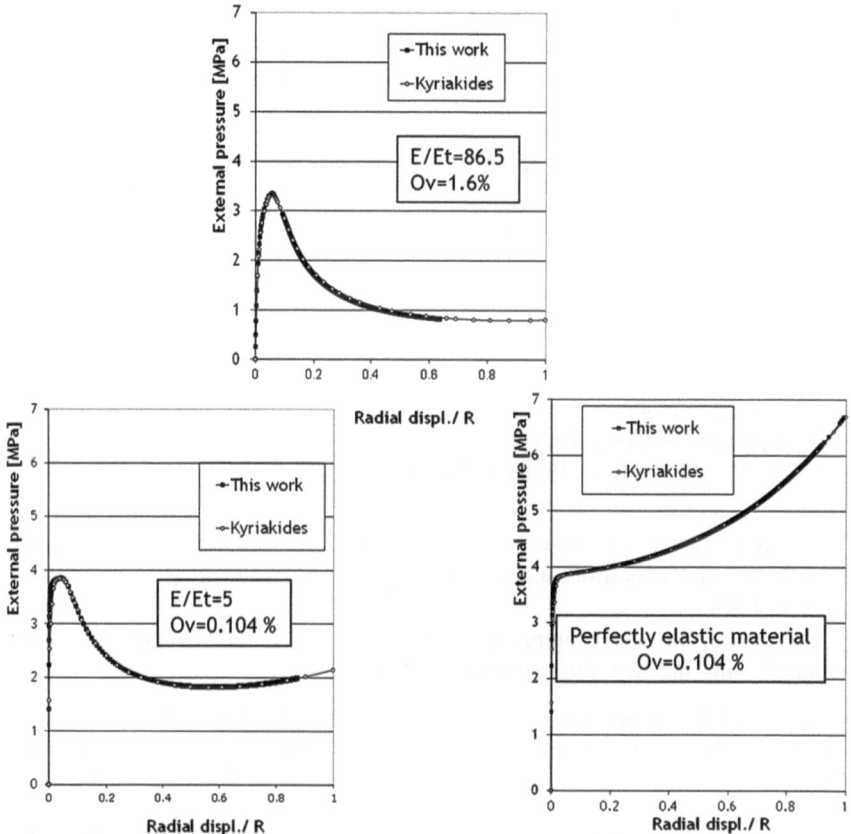

Fig. 3.8 Infinite tube. Qualification of the finite element model for the pre and post-collapse equilibrium path

In the test cases, Kyriakides adopted a bilinear stress hardening law.

Figure 3.8 shows the curves [*External pressure vs. Radial displ./R*] for a pipe with D/t = 35, D = 1.2522″, E/σ_y = 225 and σ_y = 31.5 kg/mm^2; it can be observed from the comparison of those results that our results coincide with the results published by Kyriakides, both in the pre critical and post critical regime.

3.3.3 Pipes Under Bending Plus External Pressure

In this section we investigate the pre-collapse, collapse and post-collapse behavior of steel pipes that are first bent and afterwards loaded with an increasing external pressure load up to collapse.

From the finite element results we obtain two important values: the critical external pressure (buckling initiation) and the buckling propagation pressure [22]. In order to determine the buckling propagation pressure it is necessary to make the Maxwell construction [21], using the [*Pressure vs. Enclosed area change diagram*] which is post-processed from the finite element results (Fig. 3.9). The propagation pressure is a very important property of the tubular material, since local bumps that drastically diminish the pipes collapse pressure are unavoidable (either during the pipeline construction or during its operation), it is important that the pipeline engineer designs the pipeline buckle arrestors using the proper propagation value.

To illustrate on the actual deformation pattern of the pipe sections, in Fig. 3.10 we show the results of a finite element analysis.

In Fig. 3.10 there are some distinct features to be recognized,

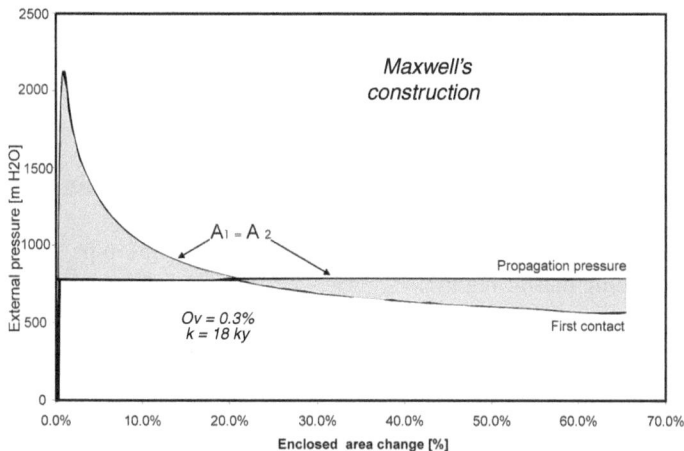

Fig. 3.9 Pressure-displacement characteristics of a long pipe under external pressure

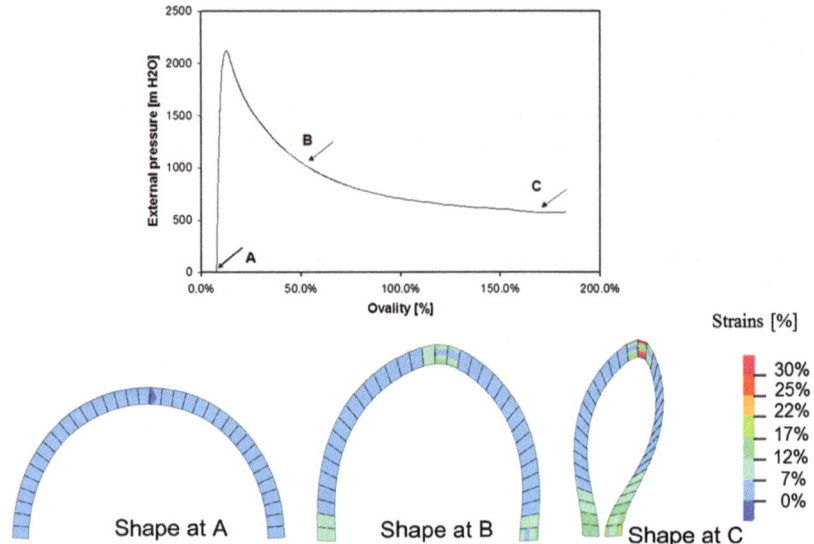

Fig. 3.10 Shape of the pipe cross section during the pre-buckling (**a**) and post-buckling regimes (**b**, **c**). 8 5/8″ OD 12.7 mm X60 pipe, initial ovality 0.3 %

- the first stage of the deformation, from $p = 0$ up to the peak pressure is the pre-buckling regime (e.g. point A, Fig. 3.10),
- the peak pressure is the collapse pressure,
- from the collapse pressure onwards, up to the final stage in which points on the inner surface of the pipe make contact among them, we have the post buckling regime (e.g. point B and C, Fig. 3.10).

3.3.3.1 Effect of the Pipe Ovality

We are going to investigate how the pipe bending curvature and its initial ovality influence the pipe critical and buckling propagation pressure.

In what follows we use the notations,

k Constant bending curvature imposed on the pipe, kept constant during the external pressure test.

k_y Value of the constant bending curvature that takes the pipe most stressed point into the plastic regime.

We analyze the behavior of an 8 5/8″ OD 12.7 mm WT X60 pipe, with a moderate initial ovality. We represent the external pressure versus the resulting ovality for each pressure level (we calculate the resulting ovality, adding to the initial values of D_{max} and D_{min} the corresponding node displacements). Figure 3.10 shows that the initial ovality is absolutely negligible when compared with the ovality induced by the loading; even the ovality corresponding to point A, that is to say the ovality induced by bending (Brazier effect [23]), is much larger than the initial ovality.

Even though the initial ovality has a strong influence on the critical collapse pressure when no bending is applied, as it was shown in the previous section, the effect of the initial ovality on the critical collapse pressure diminishes when the imposed curvature is increased. If a perfectly round tube is bent the cross section is ovalized, when the bending increases, the Brazier-ovality grows and therefore the initial ovality becomes less important as compared with this bending-induced ovality.

In Fig. 3.10 we represent the initial shape of the pipe cross section and its shape for the moment at which the inner surface closes onto itself. Please notice the strong asymmetry of the final cross section (point C).

In Fig. 3.11 we measure the applied curvature with the radius "R" and with the maximum bending strain (as a reference we indicate the radius of a typical reel used to lay marine pipelines).

The effect of the pipe initial ovality on their collapse propagation pressure is negligible for any bending situation, as shown in Fig. 3.12.

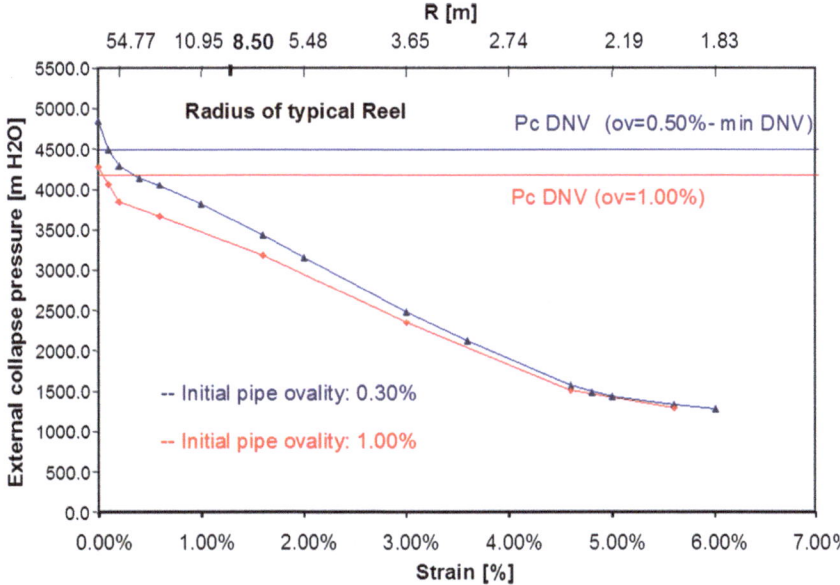

Fig. 3.11 Infinite pipe model. Ovality effect on the collapse pressure. 8 5/8" OD 12.7 mm X60 pipe

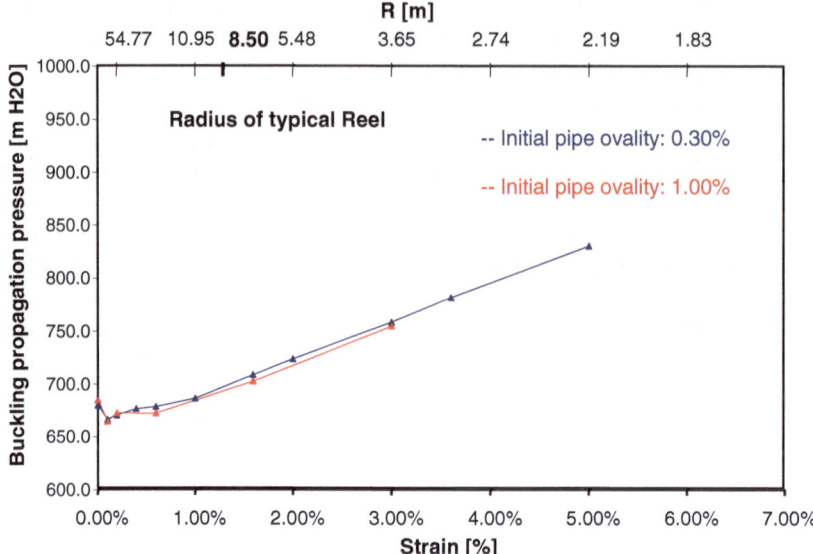

Fig. 3.12 Infinite pipe model. Ovality effect on the collapse propagation pressure. 8 5/8″ OD 12.7 mm X60 pipe

3.3.3.2 Effect of the Pipe Eccentricity

For low values of applied bending the eccentricity effect on the pipes collapse pressure is much lower than the ovality effect, and it is almost independent of the applied bending (Fig. 3.13). The eccentricity effect on the pipes collapse propagation pressure is not very relevant either (Fig. 3.14).

3.3.3.3 Effect of the Pipe Residual Stresses

In Figs. 3.15 and 3.16 we present the effect of the residual stresses on the collapse pressure and collapse propagation pressure, for various values of imposed bending (the bending is measured, in these figures, with the relation between the imposed curvature k and the curvature that yields the most strained fiber of the pipe section k_y).

The effect of the residual stresses on the external collapse pressure depends on the applied bending. For the lower values of curvature, the external collapse pressure decreases when the residual stresses absolute value increases, but for higher bending the collapse pressure increases when the residual stresses change from negative to positive values. Anyway, the effect of the residual stresses on the pipe critical collapse pressure is quite low when a strong bending is applied.

The effect of the residual stresses on the collapse propagation pressures is not very important, with or without bending.

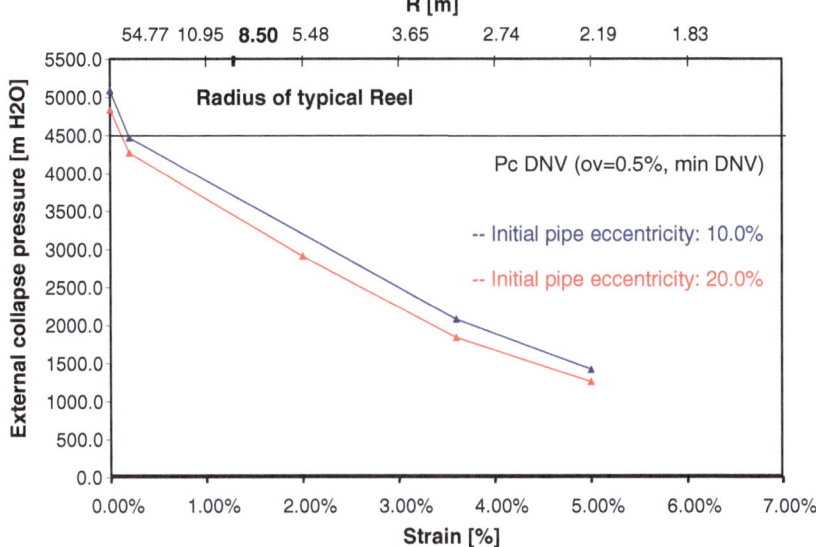

Fig. 3.13 Infinite pipe model. Eccentricity effect on the external collapse pressure. 8 5/8″ OD 12.7 mm X60 pipe

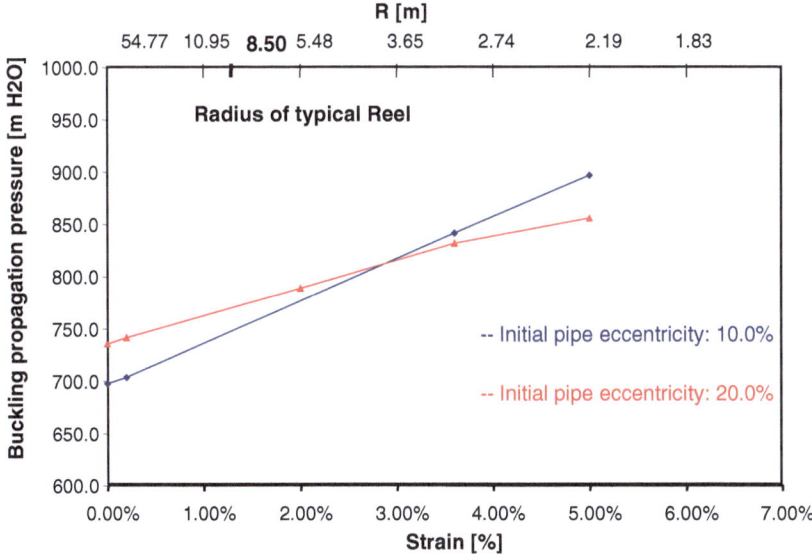

Fig. 3.14 Infinite pipe model. Eccentricity effect on the collapse propagation pressure. 8 5/8″ OD 12.7 mm X60 pipe

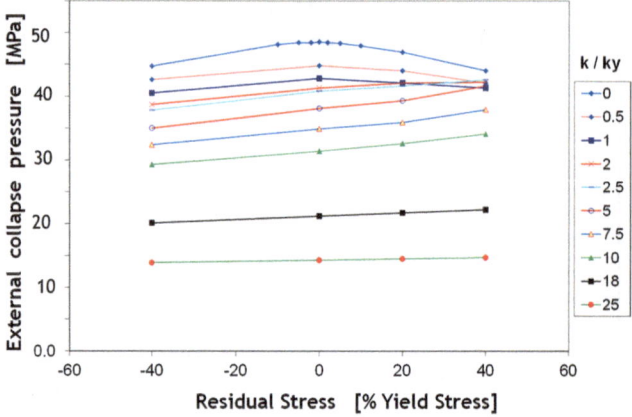

Fig. 3.15 Infinite pipe. Residual stresses effect on the external collapse pressure ($\sigma_R > 0$ indicates compression at the inner radius). 8 5/8″ OD 12.7 mm X60 pipe

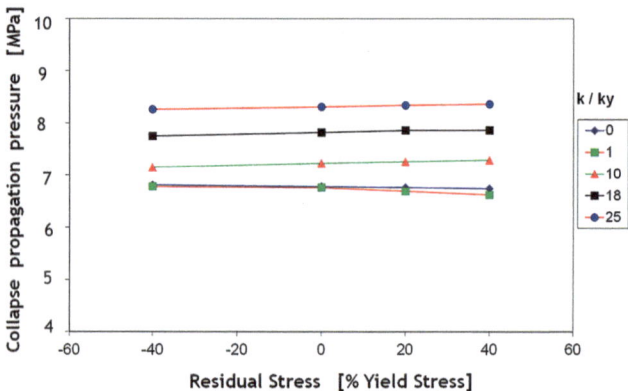

Fig. 3.16 Infinite pipe. Residual stresses effect on the collapse propagation pressure. 8 5/8″ OD 12.7 mm X60 pipe

3.3.3.4 Effect of the Imposed Bending

As it can be seen in the above figures, bending diminishes the external collapse pressure of the pipes, due to the fact that it increases its ovality ("Brazier effect").

It is also interesting to observe that bending increases the collapse propagation pressure.

3.4 Three Dimensional Finite Element Model of Finite Pipes

Following what we did above, we use a total Lagrangian formulation that incorporates the geometrical nonlinearity coming from the large displacements/rotations and the material nonlinearity coming from the elasto-plastic constitutive relation. We develop the finite element analyses using the code ADINA [3] and the MITC4 shell element.

As in the previous section, follower loads are used to model the hydrostatic external pressure and the residual stresses are modeled with a linear distribution through the thickness.

3.4.1 Residual Stresses

We use the slit ring test at the testing laboratory for measuring the residual stresses in the pipe samples [13, 24, 25]. As we see from Fig. 3.5 the correct measurement of the residual stresses is fundamental for determining the external collapse pressure of a pipe under external pressure only.

When implementing the slit-ring test, some laboratories use long slit-ring samples (L/D >3) and other laboratories use short slit-ring samples (L = 25 mm). We prefer to use long samples because they represent the averaged effect of the residual stress distribution.

3.4.1.1 Three Dimensional Simulation of the Slit-Ring Test

To check the capability of our 3D finite element models to simulate different residual stress patterns we consider a 9 5/8″ OD 47 lb/ft P110 casing with $\sigma_y = 758$ MPa and $\sigma_R = 0.2\ \sigma_y$.

We calculate the value of the samples opening in a slit-ring-test using the analytical relations between residual stresses and sample openings as well as using the 3D finite element model shown in Fig. 3.17, where the residual stresses are simulated with a linear distribution across the thickness, and the slit of the cylindrical sample is simulated by removing a row of elements.

The results are shown in Table 3.4. Hence, our 3D finite element procedure for representing the residual stresses can be considered as realistic enough.

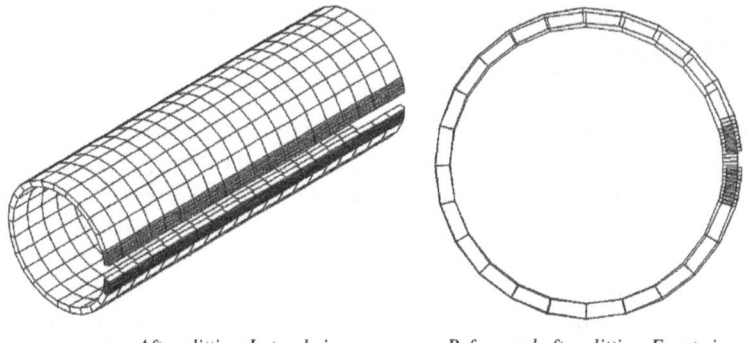

After slitting. Lateral view.   *Before and after slitting. Front view.*

Fig. 3.17 Three dimensional finite element simulation of the slit-ring test

Table 3.4 Validation of the residual stresses simulation using 3D finite element models (a is the sample opening)

Sample length	$\frac{a_{FEA}}{a_{analytical}}$
25 mm	1.02
3D	0.99

3.5 Main Observations

It was shown that the models of very long pipes do not contain enough information on the samples geometry to provide accurate predictions on the collapse pressure. However, they are very useful for developing parametric studies.

Using those models, we analyzed the pre and post-collapse regimes of very long pipes, and the effect of the initial ovality, eccentricity and residual stresses on the collapse and collapse propagation pressures.

From our numerical results we draw the following conclusions,

- the initial ovality has very important influence on the pipe collapse pressure when only external pressure is applied, but this influence diminishes as the bending curvature increases,
- the initial ovality has a negligible influence on the collapse propagation pressure under any bending condition,
- for low values of applied bending, the eccentricity effect on the collapse pressure is much lower than the ovality effect, and it is almost independent of the applied bending,
- the effect of the residual stresses on the external collapse pressure depends on the applied bending, they are a fundamental ingredient for determining collapse pressures under external pressure only, but it is quite low when a strong bending is applied,

- the effect of the residual stresses on the collapse propagation pressures is not very important, with or without bending,
- bending diminishes the external collapse pressure of the pipes, due to the fact that it increases their ovality.

References

1. Bathe K-J (1996) Finite element procedures. Prentice Hall, Saddle River
2. Hill R (1971) The mathematical theory of plasticity. Oxford University Press, New York
3. ADINA R&D. The ADINA System. Watertown
4. Dvorkin EN, Vassolo SI (1989) A quadrilateral 2-D finite element based on mixed interpolation of tensorial components. Eng Comput 6:217–224
5. Dvorkin EN, Assanelli AP (1989) Elasto-plastic analysis using a quadrilateral 2-D element based on mixed interpolation of tensorial components. In: Owen et al. DRJ (ed) Proceeding of second international conference on computational plasticity
6. Dvorkin EN, Assanelli AP, Toscano RG (1996) Performance of the QMITC element in two-dimensional elasto-plastic analyses. Comput Struct 58:1099–1129
7. Brush DO, Almroth BO (1975) Buckling of bars, plates and shells. Mc Graw Hill, New York
8. Timoshenko SP, Gere JM (1961) Theory of elastic stability. Mc Graw Hill, New York
9. Corona E, Kyriakides S (2000) An unusual mode of collapse of tubes under combined bending and pressure. ASME J Press Vesse Technol 109:302–304
10. Fowler JR, Hormberg B, Katsounas A (1990) Large scale collapse testing. SES Report, Prepared for the Offshore Supervisor Committee, American Gas Association
11. Tokimasa K, Tanaka K (1986) FEM analysis of the collapse strength of a tube. J Press Vessel Technol 108:158–164
12. Kanda M, Yazaki Y, Yamamoto K, Higashiyama H, Sato T, Inoue T et al (1983) Development of NT-series oil-country tubular good. Nippon Steel Techn Rep 21:247–262
13. Krug G (1983) Testing of casing under extreme loads. Institute of Petroleum Engineering, Technische Univesitat, Clausthal
14. Mimura H, Tamano T, Mimaki T (1987) Finite element analysis of collapse strength of casing. Nippon Steel Techn Rep 34:62–69
15. Toscano RG, Amenta PM, Dvorkin EN (2002) Enhancement of the collapse resistance of tubular products for deepwater pipeline applications. In: IBC'S Offshore Pipeline Technology, conference documentation
16. Dvorkin EN, Bathe K-J (1984) A continuum mechanics based four-node shell element for general nonlinear analysis. Eng Comput 1:77–88
17. Bathe K-J, Dvorkin EN (1985) A four-node plate bending element based on Mindlin/Reissner plate theory and a mixed interpolation. Int J Numer Methods Eng 21:367–383
18. Bathe K-J, Dvorkin EN (1986) A formulation of general shell elements—the use of mixed interpolation of tensorial components. Int J Numer Methods Eng 22:697–722
19. Bathe K-J, Dvorkin EN (1983) On the automatic solution of nonlinear finite element equations. Comput Struct 17:871–879
20. Fowler JR, Klementich EF, Chappell JF (1983) Analysis and testing of factors affecting collapse performance of casing. ASME J Energy Resour Technol 105:574–579
21. Kyriakides S (1994) Propagating instabilities in structures. Adv Appl Mech 30:67–189
22. Palmer AC, Martin JH (1975) Buckle propagation in submarine pipelines. Nature 254:46–48
23. Brazier LG (1927) On the flexure of thin cylindrical shells and other thin sections. Proc Roy Soc Lond, Math Phys A 116:104–114

24. Marlow RS (1982) Collapse performance of HC-95 casing. Report for the API PRAC Project N° 80–30
25. Tamano T, Mimaki T, Yaganimoto S (1983) A new empirical formula for collapse resistance of commercial casing. In: ASME, proceeding 2nd international offshore mechanics and Arctic engineering symposium, Houston, pp 489–495

Chapter 4
Experimental Validation of the Finite Element Models. Applications: Slotted Pipes and Axial Loads

4.1 Introduction

In previous chapters we discussed the finite element models that we use for analyzing the collapse and post-collapse behavior of steel pipes.

Since important technological conclusions are derived from the output of these models, it is of utmost importance to validate their results.

In Sect. 4.2 we describe the experimental validation program that we set up and in Sect. 4.3 we discuss the actual validation process.

In Sect. 4.4 we use our finite element models to analyze the collapse of slotted pipes and present the experimental validation of the results.

Finally, in Sect. 4.5 we use our finite element models to study the effect of axial loads on the collapse behavior of steel pipes.

4.2 The Experimental Validation Program

The test objectives were,

- determine the collapse loading and the post-collapse equilibrium path for external pressure loading,
- determine the effect of bending on the collapse strength of the pipe specimens by first applying external pressure then, while maintaining a constant external pressure, increase bending until collapse occurs ($\mathbf{P} \rightarrow \mathbf{B}$),
- determine the effect of bending on the collapse strength of one pipe specimen by first applying bending then, while maintaining a constant bending strain, increase the external pressure until collapse occurs ($\mathbf{B} \rightarrow \mathbf{P}$).

To obtain information for the model, mechanical tests on samples taken in the circumferential direction were made, the hoop residual stresses were measured using slit-ring test and the geometry of the samples was carefully acquired (see Appendix).

E. N. Dvorkin and R. G. Toscano, *Finite Element Analysis of the Collapse and Post-Collapse Behavior of Steel Pipes: Applications to the Oil Industry*, SpringerBriefs in Computational Mechanics, DOI: 10.1007/978-3-642-37361-9_4,

Therefore, the experimental work involved performing initial geometric measurements, material property tests and full-scale collapse tests (external pressure only), full-scale $P \to B$ tests and a full-scale $B \to P$ test. Nine samples were tested; all of them conforming to API 5L grade X65 [1]. The nominal dimensions for each sample are indicated in Table 4.1.

Geometrical Characterization of the Samples

The outer surface of the nine samples was mapped using the IMS (Imperfection Measuring System) [2–5] while the thickness of the samples was mapped, using a standard ultrasonic gauge. The OD Fourier mode distributions and the thickness maps are presented in the Appendix, as well as the description of the IMS.

A few comments can be made about these geometric imperfections,

- the imperfection that controls the value of the buckling pressure is the second mode of the shape Fourier decomposition, which is coincident with the first buckling mode [2],
- the value of that second mode is quite different (lower) from the ovality measured with a standard API ovalimeter [6].

Mechanical Characterization of the Samples

Coupon Tests

For each pipe sample the following determinations of the yield stress were made [7],

- coupons in the circumferential direction, tension and compression tests,
- coupons in the axial direction, tension and compression tests.

Ring Splitting Tests

Ring splitting tests were performed for each supplied pipe to measure the opening displacement of the ring sections, from which residual stress estimations were made. Residual stresses were calculated assuming a linear-elastic bending stress distribution through the wall thickness. Table 4.2 summarizes the hoop compressive yield strengths and hoop residual stresses for each specimen.

Table 4.1 Test specimens

Sample	Nominal OD (mm)	Nominal thickness (t) (mm)	$\frac{OD}{t}$	Test type
1	353	22	16.05	External pressure
2	353	22	16.05	$P \to B$
3	353	22	16.05	$P \to B$
4	323.85	17.65	18.35	Ext. pressure
5	323.85	17.65	18.35	$P \to B$
6	323.85	17.65	18.35	$P \to B$
7	323.85	20.30	15.95	Ext. pressure
8	323.85	20.30	15.95	$P \to B$
9	323.85	20.30	15.95	$B \to P$

Table 4.2 Circumferential compression yield stress and circumferential residual stresses

Sample	$\sigma_y^-(hoop)$ (MPa)	$\frac{\sigma_R(hoop)}{\sigma_y^-(hoop)}$ (%)
1	589.58	4.98
2	587.62	4.99
3	580.26	7.94
4	537.00	36.90
5	536.51	36.90
6	498.64	40.71
7	501.98	18.10
8	501.49	10.03
9	492.46	16.40

4.2.1 Full-Scale Tests

C-FER's Deepwater Experimental Chamber was used for the full-scale tests [8]. The chamber, shown in Fig. 4.1, has a tested pressure capacity of 62 MPa, with an inside diameter of 1.22 m and an overall inside length of 10.3 m.

Collapse and Buckle Propagation Tests
Three collapse and buckle propagation tests were conducted. Two of the collapse tests required pressures in excess of 62 MPa. To achieve higher pressures, a secondary pressure vessel was used inside of the Deepwater Experimental Chamber, allowing pressures up to 80 MPa (Fig. 4.2). After initial collapse, water

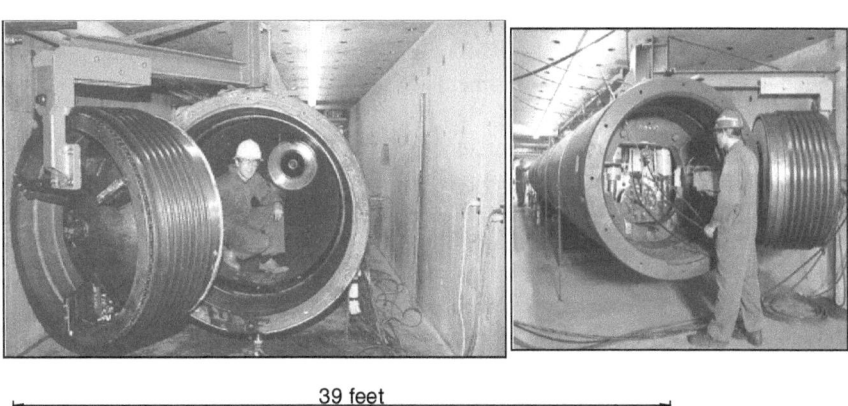

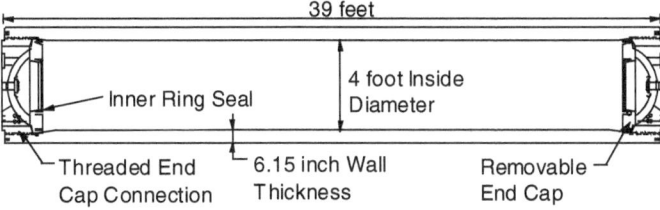

Fig. 4.1 C-FER's deepwater experimental chamber

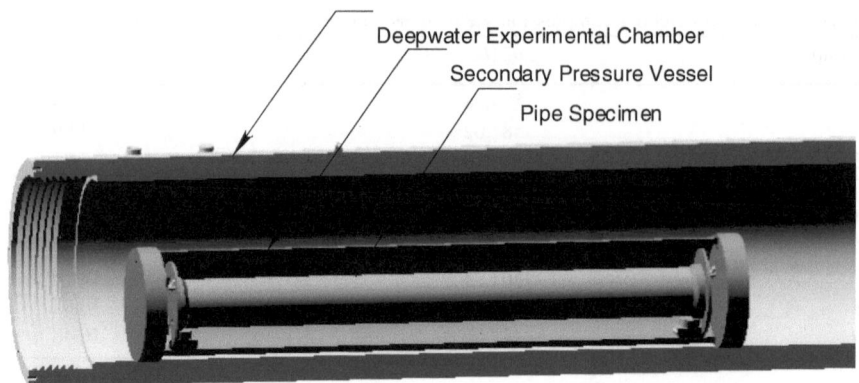

Fig. 4.2 Pipe-in-pipe set-up for high pressure tests

was continued to be pumped into the pressure vessel to propagate the collapse buckle. The propagation pressure for the three tests averaged 24 % of collapse pressure.

Measurements taken during the test included primary and secondary chamber pressures, specimen internal pressure (which was maintained near atmospheric pressure), the volume of water being pumped into the chamber and the volume of water coming out of the specimen.

P → B tests
Five **P → B** tests were performed. To perform these tests, a custom-built pipe bending system was installed inside the Deepwater Experimental Chamber (maximum bending strain of 1.5 % for a 353 mm OD pipe). The bending system applied equal and opposite end moments to the specimen ends using concentrated loads from hydraulic rams (Fig. 4.3).

Measurements for the tests included chamber pressure (specimen external pressure), specimen internal pressure, hydraulic actuator pressure, strains from four strain gauges on the test specimen, and specimen end rotation. Global bending strain was calculated based on the end rotation measurements.

Collapse of many of the specimens was characterized by an audible "*bong*", a sudden decrease in external pressure and a sudden increase in specimen internal pressure.

B → P Test
A single **B → P** test was performed. The purpose of this test was to quantify the effect of load path on the critical pressure versus bending strain relationship.

Test set-up and measurements for this test were identical to the **P → B** tests described in the previous section.

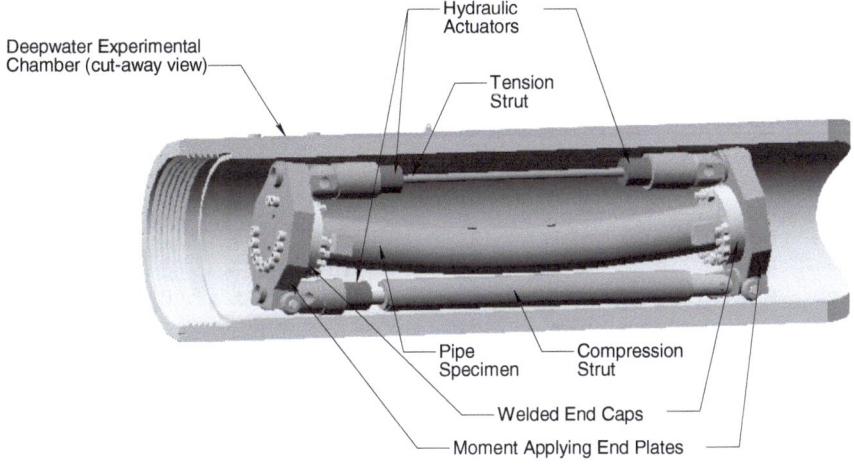

Fig. 4.3 Combined pressure and bending set-up

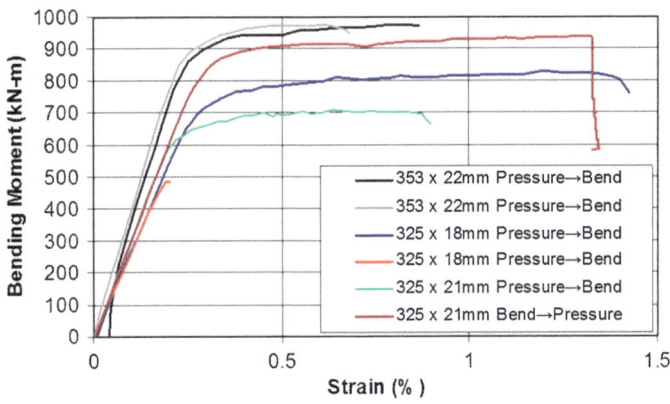

Fig. 4.4 Bending moment—strain plots

Experimental Results

Moment-strain plots for each of the bend tests are shown in Fig. 4.4.

The relation between increasing applied bending and decreasing collapse pressure agrees with the same tendency reported in Chap. 3 as a result of finite element simulations of $\mathbf{B} \to \mathbf{P}$ tests.

Results of the full-scale tests are presented in Table 4.3 and shown graphically in Fig. 4.5.

The tests successfully demonstrated the influence of bending on collapse strength for the specimens tested. Bending diminishes the external collapse pressure of the pipes, due to the fact that it increases its ovality ("Brazier effect") and introduces a biaxial state of stress. In addition, the stability of the pipe cross-

Table 4.3 Full scale test results

Sample	Test type	Collapse pressure (MPa)	Propagation pressure (MPa)	External pressure (MPa)	Peak moment (kN-m)	Strain at peak moment (%)	Average bending strain (%)
1	External pressure	80.3	18.4	–	–	–	–
2	P → B	–	–	57.2	977	0.79	–
3	P → B	–	–	51.1	975	0.61	–
4	External pressure	57.3	13.2	–	–	–	–
5	P → B	–	–	31.1	827	1.18	–
6	P → B	–	–	46.9	485	0.19	–
7	External pressure	63.9	16.6	–	–	–	–
8	P → B	–	–	50.4	709	0.64	–
9	B → P	52.3	–	–	938	–	1.33

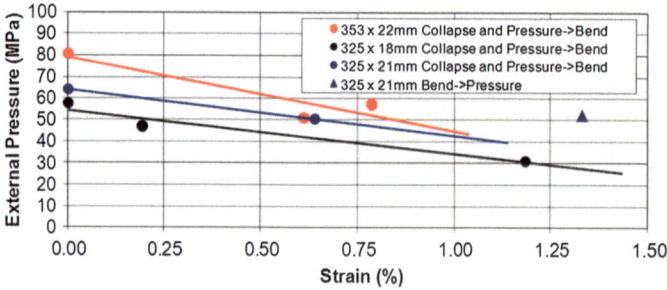

Fig. 4.5 Full-scale test results

section is dependent on the sequence of load application, as evidenced by the single **B → P** test result, which was approximately 50 % higher than would have been expected for a test conducted with a **P → B** load path. The dependency of collapse on the sequence of loading is related to the loading/unloading sequences (and resulting material stiffness changes) that arise around the circumference of the pipe cross section. The dependency of collapse on the load path is also discussed in [9] for **B → P** and **P → B**.

All the collapsed samples were visually inspected for cracks and despite the very large strains that were developed in the post-collapse regime [10], no cracks were found. This demonstrates the high ductility of the steel pipes. One advantage of this ductility is that, in the unlikely event a buckle is formed, the chance of a wet-buckle[1] [11] is greatly reduced.

---

[1] We call a wet buckle when as a result of the collapse, water goes into the pipeline.

4.3 Validation of the Finite Element Results

The 3D finite element models of pipes were developed using a material and geometrical nonlinear formulation [12] and they incorporate the following features,

- geometry, as described by the OD mapping and by the thickness distribution,
- MITC4 shell element [13–15],
- von Mises elastic-perfectly plastic material model with the yield stress corresponding to the sample's hoop yield stress in compression. In this model, the plastic anisotropy of the material was neglected,
- circumferential residual stresses (experimentally measured),
- contact elements on the pipe inner surface [12] in order to prevent its interpenetration in the post collapse regime,
- the nonlinear equilibrium path was tracked using the algorithm described in Ref. [16],
- the boundary conditions depend on the collapse chamber used in each case.

The ADINA [17] code is used for the analyses.

4.3.1 Numerical Results

Because the characteristics of the collapse chambers used in these tests, shown in Figs. 4.2 and 4.3, the external pressure acts on the lateral surface of the pipes and also it introduces an axial compression on them.

In what follows, in order to validate the numerical models, for the nine tests described in Table 4.1 we compare the finite element results with the full-scale results.

Sensitivity of the Numerical Results
It is important to realize that the laboratory determined diameter and wall thickness of the pipe samples is subjected to the normal indeterminations of lab measurements [18].

We also characterize the mechanical properties, such as the yield stress and the residual stresses, with a constant value although they have a degree of variability inside the sample.

Hence, we need to be able to evaluate the sensitivity of the numerical results to small changes in the data.

The sensitivity of the collapse pressure to the yield stress value can be written as,

$$\frac{\partial p_{col}}{\partial \sigma_y} = \frac{p_{col}\left(\sigma_y^0 + \Delta\sigma_y; \sigma_{Res}^{Hoop}; \sigma_{Res}^{Axial}\right) - p_{col}\left(\sigma_y^0; \sigma_{Res}^{Hoop}; \sigma_{Res}^{Axial}\right)}{\Delta\sigma_y}$$

In the above equation,

σ_y^0 base value for the yield stress.

$\Delta\sigma_y$ admissible variation for the yield stress.

σ_{Res}^{Hoop} residual stress value in the hoop direction.

σ_{Res}^{Axial} residual stress value in the axial direction.

Initial baseline analyses were performed using the above-discussed geometrical and mechanical properties. Axial residual stresses were not measured in the test program and were assumed to be zero for this baseline case. However, to estimate the sensitivity of collapse pressure to axial residual stresses, we assumed the axial residual stresses to have a linear distribution through the pipe wall thickness. A maximum value was assumed when the residual stress equaled the absolute value of the measured hoop residual stresses and they were tensile on the outer fibers. A minimum value was assumed when they were equal to the absolute value of the hoop residual stresses and they were compressive on the outer fibers. A variation of ±10 % was also assumed in the values of yield stress and hoop residual stress to estimate the sensitivity of collapse to changes in their values.

The sensitivity of the collapse pressure to the value of the hoop residual stresses is,

$$\frac{\partial p_{col}}{\partial \sigma_{Res}^{Hoop}} = \frac{p_{col}\left(\sigma_y^0;\, \sigma_{Res}^{Hoop} + \Delta\sigma_{Res}^{Hoop};\, \sigma_{Res}^{Axial}\right) - p_{col}\left(\sigma_y^0;\, \sigma_{Res}^{Hoop};\, \sigma_{Res}^{Axial}\right)}{\Delta\sigma_y^{Hoop}}$$

The sensitivity of the collapse pressure to the value of the axial residual stresses can be calculated using a similar procedure.

External Pressure Only (Samples 4, 1 and 7)
Sample 4
The numerical results compared with the experimental ones are shown in Fig. 4.6.

The baseline finite element result presents an excellent agreement with the experimental result. The difference of 3.4 % between the finite element predicted and the experimentally determined collapse pressures can be attributed to the non-homogeneous yield stress and to the indetermination in the value of the axial residual stresses, as can be seen in Fig. 4.6. In Fig. 4.7 we compare the experimentally and numerically determined (*External Pressure vs. Internal Volume Reduction*) diagrams. In this figure we also compare the deformed meshes corresponding to the post-collapse regime with the collapsed pipe profile. It is important to realize that the numerical analysis was halted before the collapse could propagate through the entire sample.

Both diagrams are practically coincident, except in the interval that goes from immediately after the pipe collapse to the point at which the experimentally and numerically determined curves merge again. In the experimental test, after collapse the chamber is abruptly depressurized and water must be pumped to regain pressure. Hence, the (*External Pressure vs. Internal Volume Reduction*) experimental path is different from the numerical one, which better represents the undersea conditions.

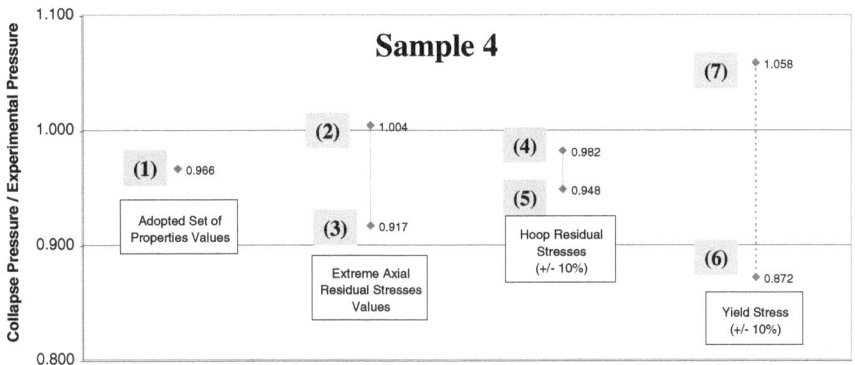

(1) Baseline case: the results were obtained using the measured data with no axial residual stresses
(2) Axial residual stressese qual to –(measured hoop residual stresses)
(3) Axial residual stressese qual to (measured hoop residual stresses)
(4) Hoop residual stresses = 0.9 baseline residual stresses
(5) Hoop residual stresses = 1.1*baseline residual stresses
(6) Yieldstress= 0.9*baseline yield stress
(7) Yieldstress= 1.1*baseline yield stress

Fig. 4.6 Sample 4: sensitivity analysis for the external collapse pressure

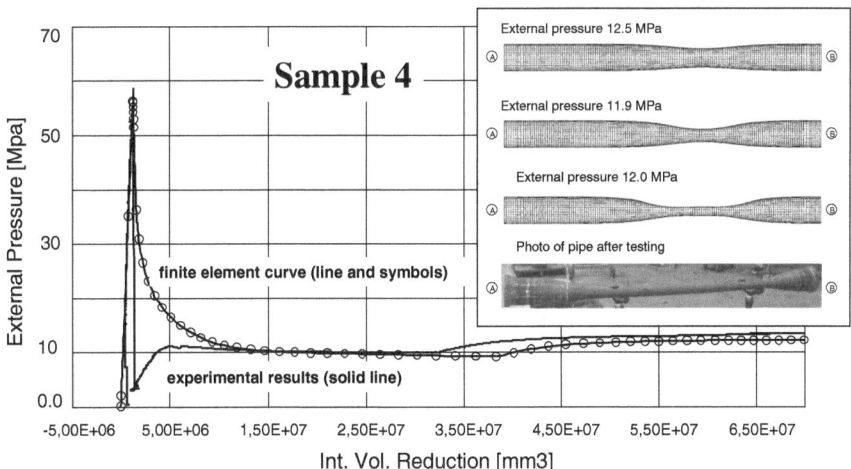

Fig. 4.7 Sample 4: External pressure versus internal volume reduction; finite element curve and experimental results

Figure 4.8 presents the deformed finite element mesh corresponding to a certain point of collapse propagation.

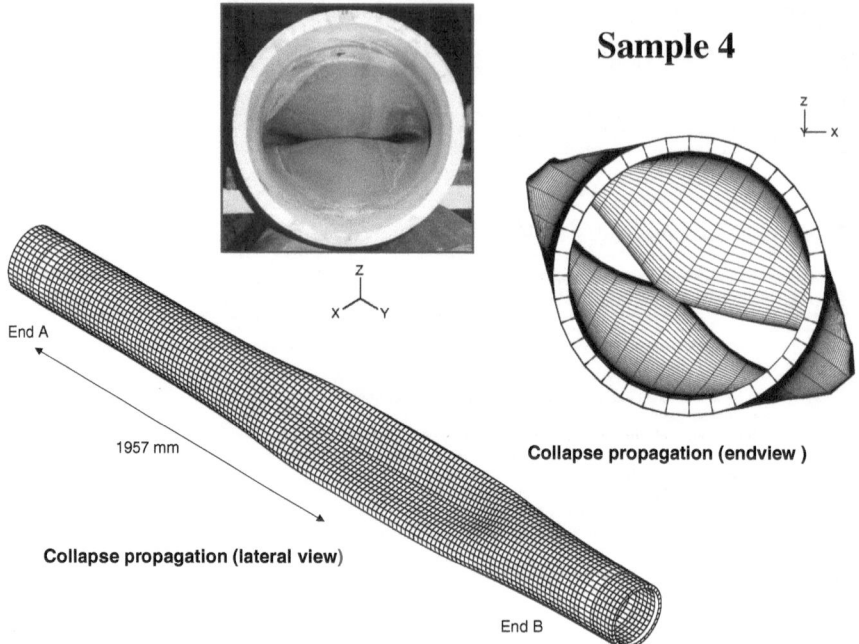

Fig. 4.8 Sample 4: Post-collapse: isometric and end view

Hence, we can assess that the post-collapse response of the finite element model, specifically the path in which the collapse propagates, has an excellent match with the experimental results.

In the post-collapse regime the pressure raises from 10 MPa to approximately 12.2 MPa (propagation pressure), due to the stiffening effect of the contact between opposite points on the inner pipe surface. The ratio between the numerical collapse propagation pressure and the experimental one is 0.89.

In Fig. 4.9 we can observe the propagation of the contact pressure along the contact line in the raising part of the equilibrium path.

It is important to note that the sensitivity of the propagation pressure to the yield stress value is quite low, in the analyzed case it is ten times lower than the sensitivity of the collapse pressure to the yield stress value.

With the finite element model, for sample 4, we obtained,

$$\frac{\partial p_{col}}{\partial \sigma_y} = 0.100$$

$$\frac{\partial p_{prop}}{\partial \sigma_y} = 0.013$$

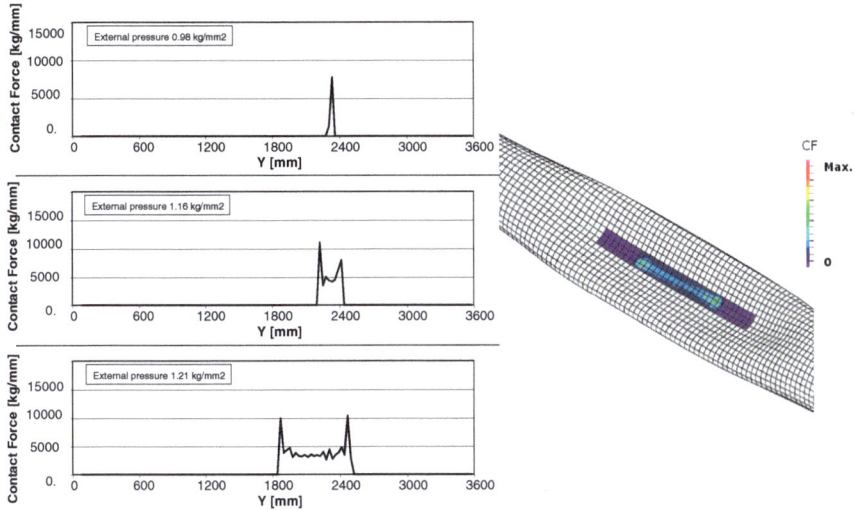

Fig. 4.9 Sample 4: Propagation of the contact pressure along the contact line

Samples 7 and 1

Figures 4.10 and 4.11 summarize the comparison between the finite element and experimental results for samples 7 and 1, respectively, while Figs. 4.12 and 4.13 present the external collapse pressure sensitivity analyses performed on both samples. For sample 7 we have also analyzed the sensitivity of the collapse pressure to the measured thickness. Since we measured the wall thickness using a manual ultrasonic device, errors in this measurement due to a wrong positioning of the ultrasonic gauge are very easy to have.

Regarding the collapse propagation pressure, the ratio between the numerical result and the experimental one is 0.99 for sample 7 and 0.87 for sample 1.

In Fig. 4.14 we map the equivalent logarithmic plastic strains. As we discussed above these strains are rather large, however these samples did not present any crack when it was visually inspected at C-FER after the collapse test.

FEA for the B → P Test

In this test, the post-collapse regime was not investigated; hence, only the collapse external pressure could be compared.

The finite element result presents an excellent agreement with the experimental result: a difference of only 3.6 % between the FEA prediction and the experimentally determined collapse pressures. This can be attributed to the non-homogeneous yield stress and to the uncertainty of the value of the axial residual stresses.

Figure 4.15 presents the curve (*Bending moment vs. average bending strain*) for sample 9.

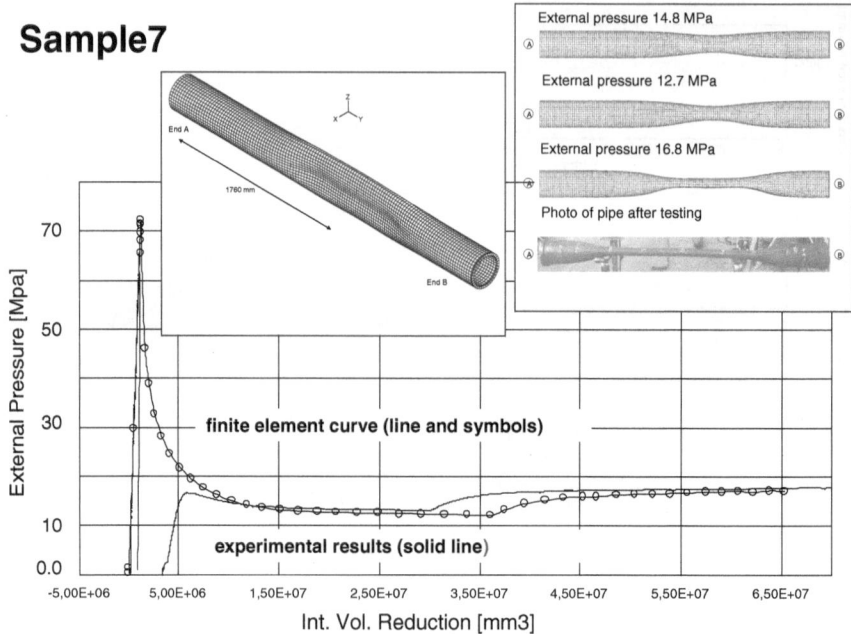

Fig. 4.10 Sample 7: External pressure versus internal volume reduction: finite element curve and experimental results

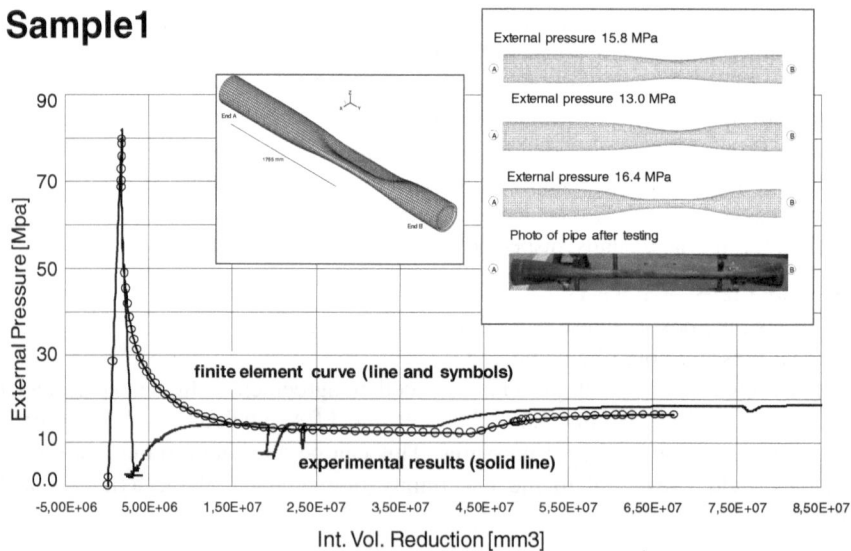

Fig. 4.11 Sample 1: External pressure versus internal volume reduction: finite element curve and experimental results

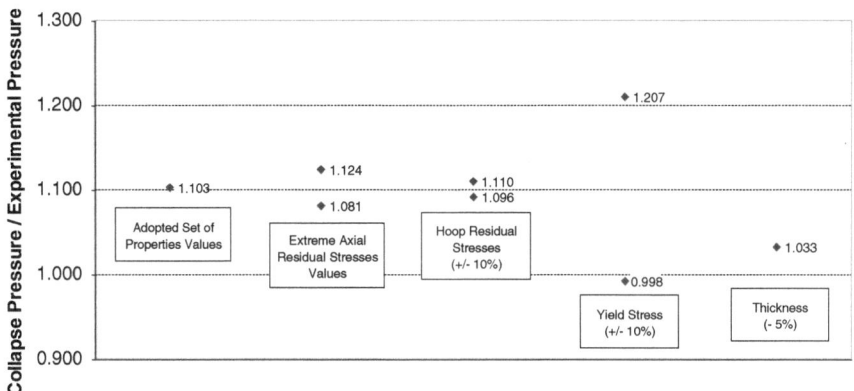

Fig. 4.12 Sample 7: Sensitivity analysis for the external collapse pressure

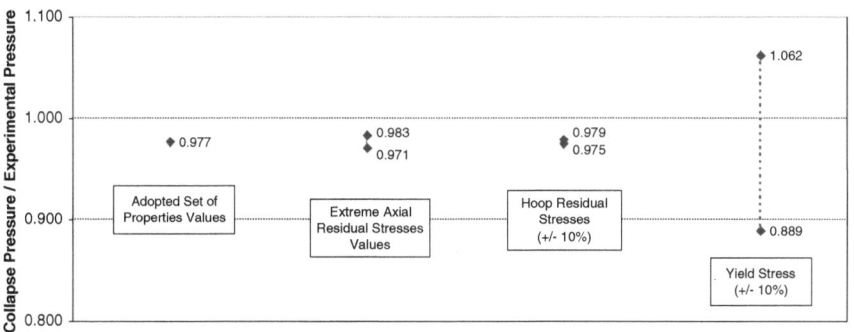

Fig. 4.13 Sample 1: Sensitivity analysis for the external collapse pressure

FEA for the P → B Test

In this experimental test, the post-collapse regime was not investigated; hence, only the collapse bending moment could be compared. Five samples were first loaded with external pressure and afterwards, maintaining constant the external pressure, they were bent up to collapse. Table 4.4 summarizes the comparison between the numerical and experimental results.

The finite element results are in excellent agreement with the experimental ones. Figures 4.16 and 4.17 present the curves (*Bending moment vs. average bending strain*) and pictures of the pipes after collapse (samples 6 and 8, respectively); Fig. 4.17 also shows the equivalent plastic strain distribution.

The sensitivity analysis results for all the samples are listed in Table 4.5.

Collapse Mode

For the pipe specimens subjected to external pressure only, the numerically predicted collapse mode matched the experimentally observed mode in only one case (Fig. 4.18). However, in the cases that included bending, the agreement was excellent (Fig. 4.19).

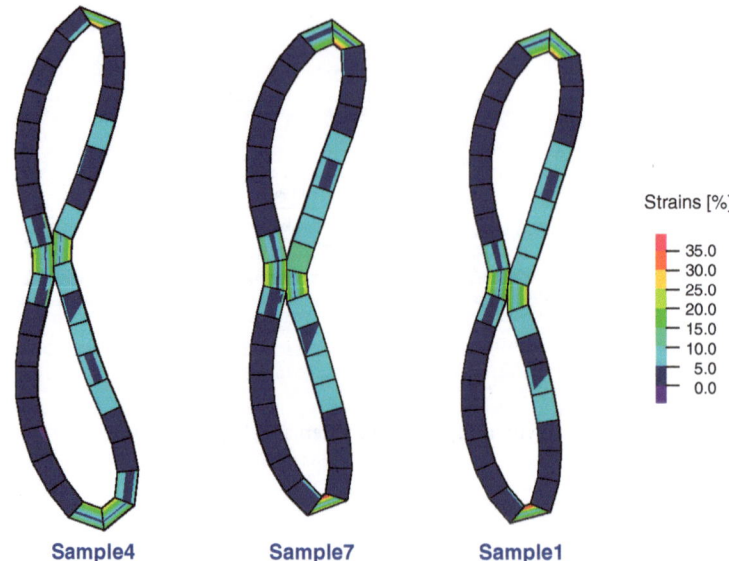

Fig. 4.14 Sample 4, 7 and 1: Hencky equivalent plastic strains

4.4 Application: Validation for Slotted Pipes

Steam Assisted Gravity Drainage (SAGD) is an oil recovery technology for producing heavy crude oil using slotted pipes. A pair of horizontal wells is drilled into the oil reservoir separated by a vertical distance of a few meters. Steam is injected into the upper well to heat the oil; then the oil, with its viscosity reduced, is collected for the lower wellbore through a slotted liner.

A 3D finite element model was developed to assess the influence of the slots on the structural resistance of the pipes.

The objective of the work described in this Section was to validate the 3D finite element model for external collapse pressure with axial compression load, by comparing the numerical results with the experimental ones. The tests were conducted at C-FER Technologies (C-FER), Canada. Two different slotting geometries were tested under different loading conditions (external pressure plus axial compression up to collapse), representative of installation and service conditions in SAGD wells.

The external collapse pressure of un-slotted pipes, with the same geometry and material than the slotted ones, are also calculated in order to evaluate the influence of the slots in the structural behavior of the pipes.

General Characteristics of the Slotted Pipes

We analyze staggered slotting patterns, and compare the structural behavior of pipes with 180, 100 and 60 spf slot (slots per foot).

Figure 4.20 shows a drawing of the samples.

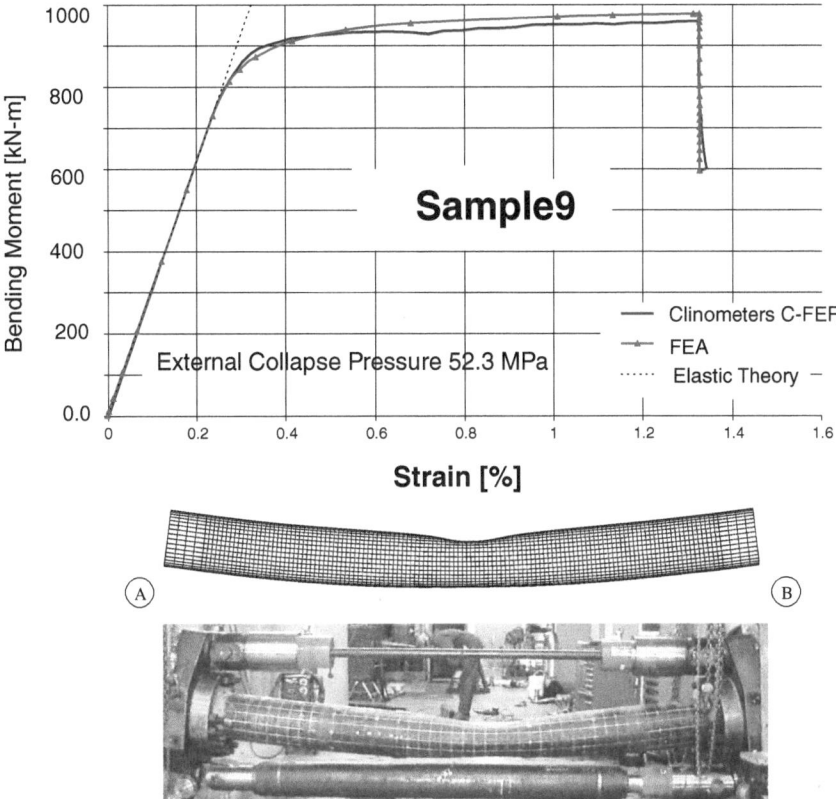

Fig. 4.15 Sample 9: Bending moment versus average bending strain

Table 4.4 Circumferential Results for pressure plus bending	Sample	$\frac{Mc_{Numerical}}{Mc_{Experimental}}$
	2	1.047
	3	1.088
	5	0.972
	6	0.998
	8	0.998

Table 4.6 summarizes the main characteristics of the slots while Table 4.7 shows the configuration of each sample.

Figure 4.21 shows a detail of the staggered pattern; "a" is equal to 0.49" (12.4 mm) for slot density #1, whereas for slot density #2 the slot separation is 1.47" (37.2 mm).

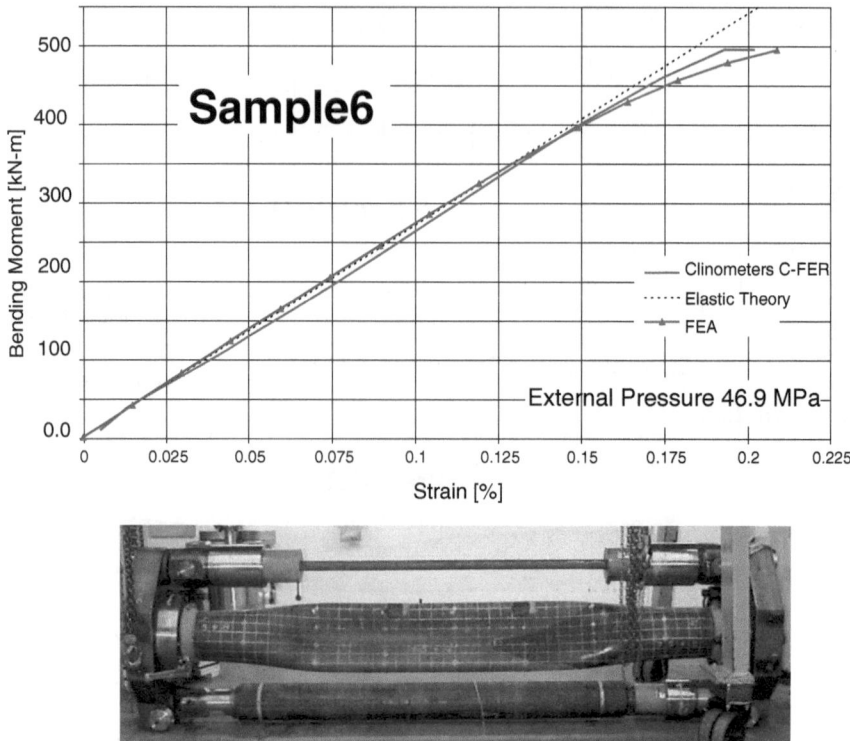

Fig. 4.16 Sample 6: Bending moment versus average bending strain and pipe after collapse

4.4.1 Numerical Model

The 3D model presents the same characteristics that were described in Sect. 4.3, but in this case we include linear gap elements to avoid the interpenetration of the slot walls.

Figure 4.22 shows an example of the finite element mesh used for the analysis of pipes with slot density #1 and a detail view of the slots.

Material Model Curves
Averaged engineering strain–stress curves were considered, taking into account two measured responses per specimen. As an example, the following curve shows the material model used for Sample #1 (Fig. 4.23).

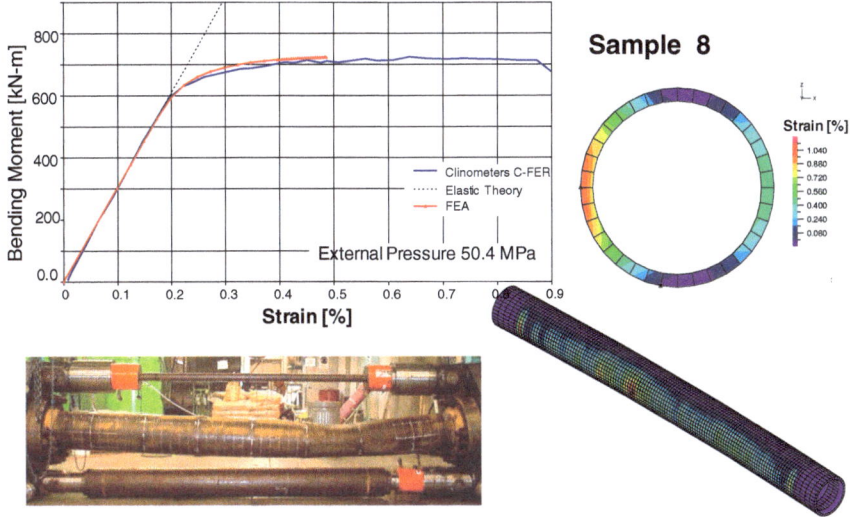

Fig. 4.17 Sample 8: Bending moment versus average bending strain and pipe after collapse. Equivalent plastic strains

Table 4.5 Sensitivity Analysis Results

	$\frac{Pc_{FEA}}{Pc_{Exp}}$	$\frac{Pc_{FEA}}{Pc_{Exp}}$	$\frac{Pc_{FEA}}{Pc_{Exp}}$	$\frac{Pc_{FEA}}{Pc_{Exp}}$	$\frac{Mc_{FEA}}{Mc_{Exp}}$	$\frac{Mc_{FEA}}{Mc_{Exp}}$	$\frac{Mc_{FEA}}{Mc_{Exp}}$	$\frac{Mc_{FEA}}{Mc_{Exp}}$	$\frac{Mc_{FEA}}{Mc_{Exp}}$
Sample	4	7	1	9	6	8	5	2	3
Baseline	0.966	1.103	0.977	0.964	0.998	0.998	0.972	1.047	1.088
Min.axial σ_R	1.004	1.124	0.983	0.956	1.048	0.998	0.972	1.048	1.088
Max.axial σ_R	0.917	1.081	0.971	0.974	0.688	0.998	0.970	1.047	1.088
Min.hoop σ_R	0.982	1.110	0.979	0.962	1.046	0.998	0.972	1.048	1.088
Max.hoop σ_R	0.948	1.096	0.975	0.968	0.902	0.998	0.972	1.047	1.089
Min. σ_y	0.872	0.998	0.889	0.903	0.166	0.827	0.854	0.843	0.919
Max. σ_y	1.058	1.207	1.062	1.022	1.336	1.156	1.088	1.223	1.247

4.4.2 Numerical Results

Figures 4.24 and 4.25 show, for Sample #1, the experimental curves (*External pressure & Axial Compression load vs. Pumped Volume*) and the numerical ones (*External pressure vs. Displacement*), where Displacement refers to the maximum diameter increase. We also include the numerical curve obtained for a pipe with the same geometry, material and axial load, but without slots. Figure 4.26 shows the effective plastic strains and hoop stresses after collapse for Sample #1. For all the cases under study, the distribution of plastic strains and stresses is similar. Finally, we calculate the width variation of the slots during the numerical tests.

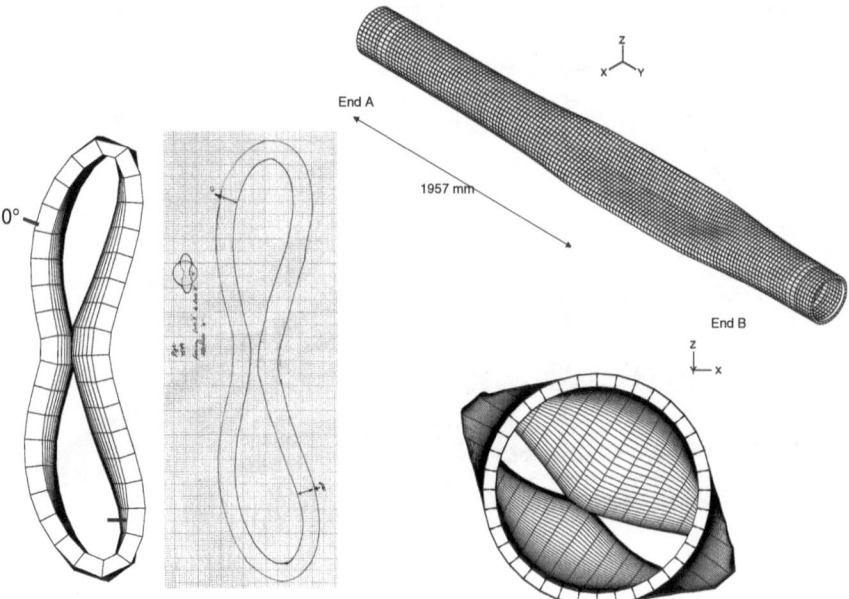

Fig. 4.18 Sample 7: .FEA and experimental predictions for the collapsed section (external pressure only)

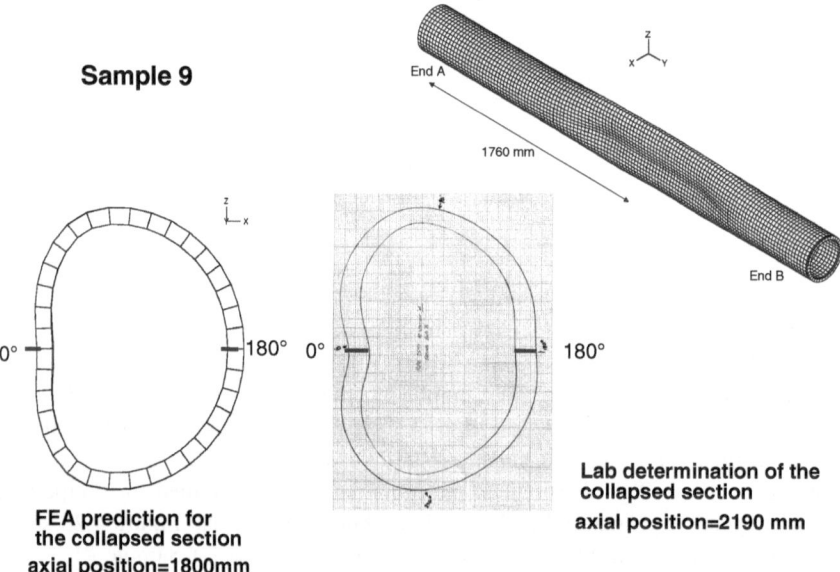

Fig. 4.19 Sample 9. Numerically predicted and experimentally observed collapse mode for the **B → P** test

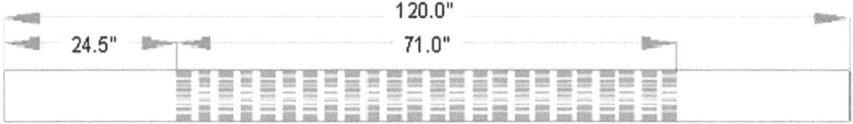

Fig. 4.20 Slotted pipes

Table 4.6 Description of the slots. Shapes #1 and #2 represent two different cross sections of the slots

	Slot Density #1 Shape #1	Slot density #1 Shape #2	Slot Density #2 Shape #1
Configuration	180 slots/ft (45 slots per column, columns spaced 3 inches apart)	180 slots/ft (45 slots per column, columns spaced 3 inches apart)	60 slots/ft (15 slots per column, columns spaced 3 inches apart)
Cut slot gap	0.020″ (0.508 mm)	0.020″ (0.508 mm)	0.020″ (0.508 mm)
Slot length on ID	0.85″ (21.59 mm)	0.85″ (21.59 mm)	0.85″ (21.59 mm)
Slot length on OD	2.1″ (53.34 mm)	2.1″ (53.34 mm)	2.1″ (53.34 mm)
Columns	24	24	24
Percentage of open area (%)	1.16	0.70	0.39

Table 4.7 Description of the samples

Sample	Slot density
1	Slot #2, shape #1
2	Slot #1, shape #1
3	Slot #1, shape #1
4	Slot #1, shape #2
5	Slot #1, shape #2

Gap Variation of the Slots

Figure 4.27 shows the curve (*Slot gap vs. external pressure*) calculated numerically. The slot gap values belong to the most deformed slot of the pipe. Even though the gap of the slots diminishes along the collapse test, the slots continue being open.

Summary

Table 4.8 summarizes the obtained results.

We can observe that the ratio between numerical versus experimental results is very good, therefore we can say that the model is validated for external pressure plus axial compression load.

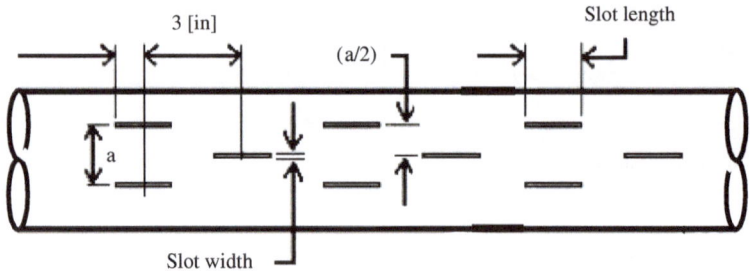

Fig. 4.21 Detail of the staggered pattern

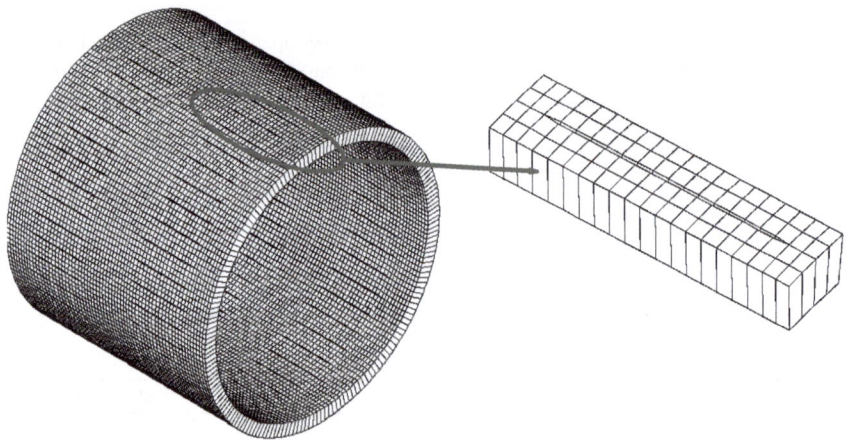

Fig. 4.22 Slotted pipe mesh. Slot detail

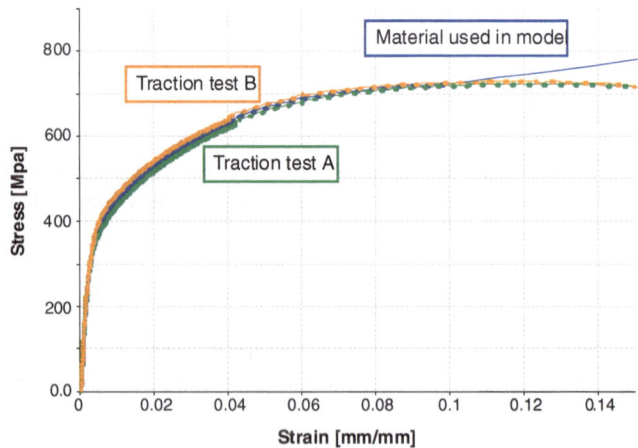

Fig. 4.23 Material model (Sample #1)

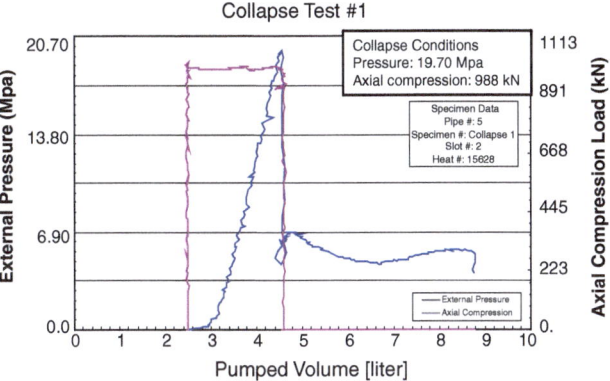

Fig. 4.24 Experimental curve (Sample #1)

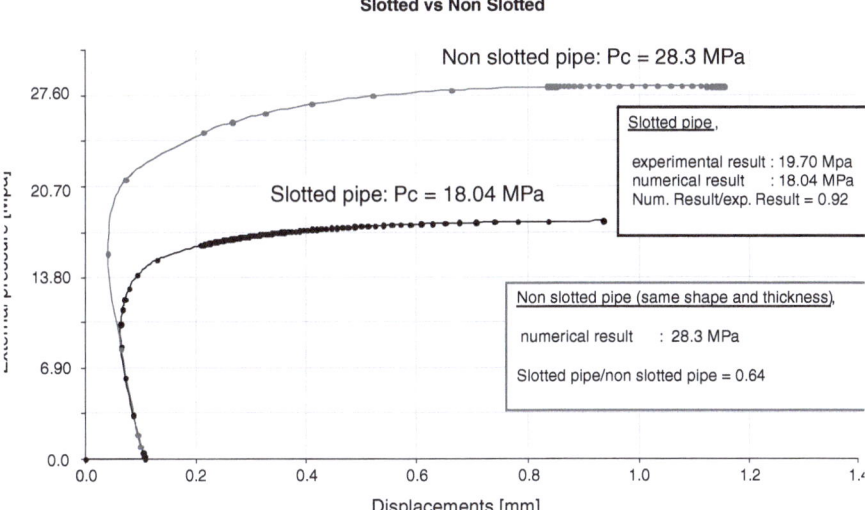

Fig. 4.25 Numerical curves (Sample #1)

Regarding the influence of the slots on the structural behavior of the pipes, the ratio between the numerical collapse pressure for both, slotted and non slotted pipes, indicates that the effect of the slots is very detrimental.

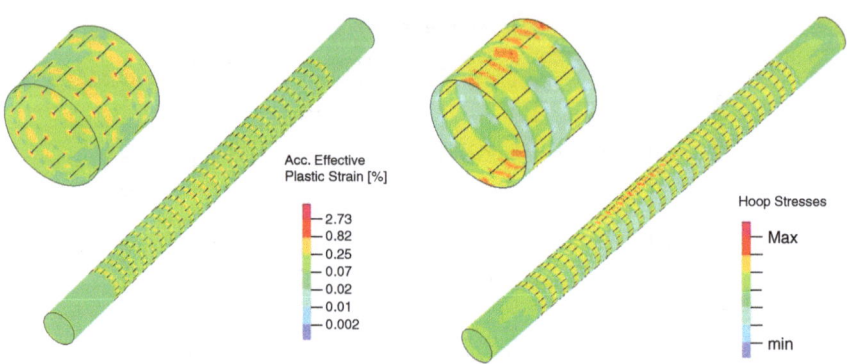

Fig. 4.26 Plastic strains and hoop stresses after collapse (Sample #1)

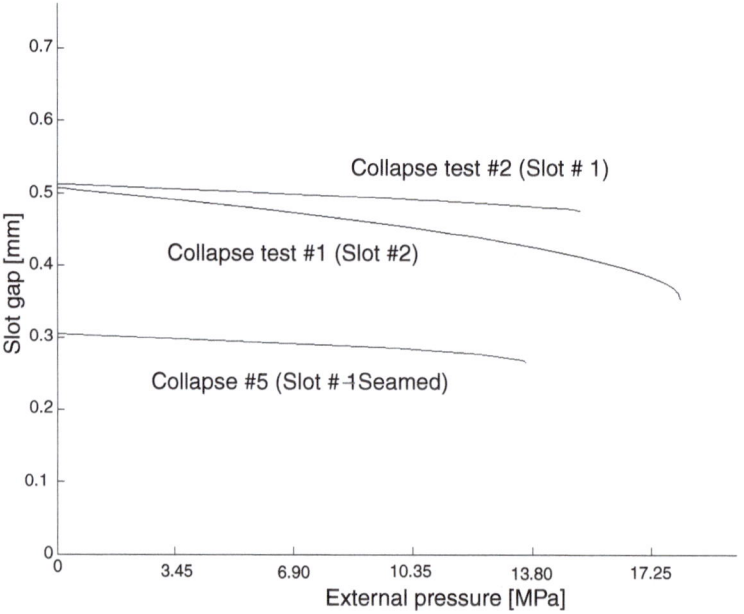

Fig. 4.27 Slot width variation versus external pressure

4.5 Application: Collapse of Steel Pipes Under External Pressure and Axial Tension

It is well known that, when steel pipes are subjected to axial tensile loads, their external collapse pressure diminishes.

Table 4.8 Summary of results

Sample	Slots per column	Experimental collapse pressure (Mpa)	Numerical collapse pressure (Mpa)	Numerical collapse pressure/ experimental collapse pressure	Numerical collapse pressure (without slots) (Mpa)	Numerical collapse pressure slotted/Numerical collapse pressure non slotted
1	15	19.70	18.04	0.92	28.30	0.64
2	45	14.45	15.16	1.05	26.75	0.57
3	45	14.45	14.2	0.98	25.53	0.56
4	45	14.33	13.6	0.95	24.24	0.56
5	45	14.39	13.6	0.95	25.06	0.54

Fig. 4.28 Grade 55 with the axial load being kept constant during the pressure loading

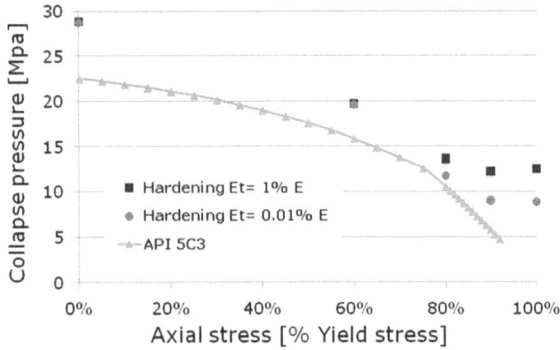

When calculating the collapse pressure of casings using the standard API 5C3 [19], even though the formulas are not applicable for axial tensile stresses up to the material yield stress, it is clear that the predicted collapse pressure tends to zero when the applied axial tensile stress tends to the material yield stress (see Figs. 4.28, 4.29 and 4.30).

When calculating the collapse pressure of submarine pipeline systems using the standard DNV OS-F101 [20] the external collapse pressure is zero when the applied axial tensile stress equals the material yield stress.

But it has been observed that even when the applied axial tensile stress matches the material yield stress, there is still a remaining capacity in the pipes for carrying external pressure [21].

In this section we investigate the above assertion and quantify, using finite element models, the collapse of steel pipes that are first subjected to axial tensile load and afterwards to external pressure.

The finite element analyses that we present here confirm that, even when the applied axial tensile load matches the material yield load, there still remains a non negligible capacity in the pipes for carrying external pressure [22].

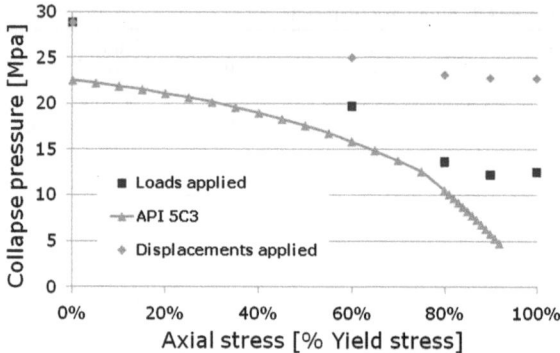

Fig. 4.29 Grade 55 with the axial displacement being kept constant during the pressure loading

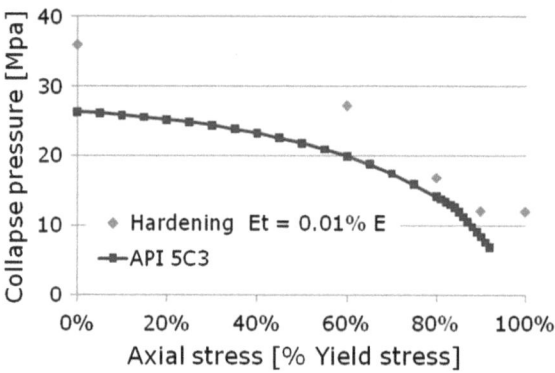

Fig. 4.30 Grade 80 with the axial load being kept constant during the pressure loading

Numerical Analyses

We consider samples with a ratio between its length (L) and its diameter (D) equal to 10, and two different load cases:

- the sample is loaded into axial tension and afterwards, *keeping constant the applied axial load*, it is loaded with external pressure up to collapse;
- the sample is loaded into axial tension and afterwards, *keeping constant the axial displacement of the sample ends*, it is loaded with external pressure up to collapse.

Even though the hardening of the material does not have any influence on the external collapse pressure of pipes when the applied load is only external pressure [2], in this case *[axial tensile load + external pressure]* we also investigate the effect of the material hardening (E_t) on the results.

For our analyses we consider a 7″ 23# pipe geometry and two material grades, Grade 55 and Grade 80. We include in our models a constant ovality of 0.5 %.

The main characteristics of the model are similar to those described in Sect. 4.3. We consider a simple bilinear material model with hardening modulus E_t.

Regarding the boundary conditions, one of the sample ends is modeled as fixed and the other one is given one axial degree of freedom.

Collapse of the 7″ 23# Grade 55

In these analyses we consider two hardening cases,

- $E_t = E/100$
- $E_t = E/10,000$.

In the above, E is the steel Young's modulus.

In Fig. 4.28 we show the results for the case in which the axial load is kept constant during the external pressure loading.

From the results in Fig. 4.28 it is clear that for the analyzed case, and when both hardenings are considered, the pipe remains with a non negligible capacity for carrying external pressure after it yields in axial tension.

Also that figure shows that the hardening effect is only significant for axial loads close to the yield load.

In Fig. 4.29 we show the results for the case in which the displacement of the sample ends are kept constant during the external pressure loading.

The results in the last figure show that when the axial end displacements are kept constant the collapse pressure is higher than when the axial load is kept constant.

Collapse of the 7″ 23# Grade 80

In Fig. 4.30 we study the effect of the axial load for a Grade 80 pipe considering only the more detrimental cases: the smallest hardening and the axial load being kept constant during the pressure loading.

The results for this case show the same tendency as the ones that we observed for the Grade 55 case.

Comments

The finite element analyses that we discussed in this Section confirm that even when the applied axial tensile load matches the material yield load there still remains a non negligible capacity in the pipes for carrying external pressure. The existing standards do not show this extra collapse strength and therefore underestimate the pipes collapse resistance.

The case in which the axial load is kept constant during the pipe external pressurization has a larger detrimental effect on the pipes external collapse pressure than the case in which the axial displacement is kept constant.

4.6   Main Observations

The agreement between the finite element predictions and the laboratory observations, both in the pre- and post-collapse regimes was very good. Therefore, the finite element models can be used as a reliable engineering tool for analyzing the effect of different imperfections, and of residual stresses, on the collapse and collapse propagation pressure of steel pipes.

The tests not only confirm the conclusions drawn in the previous Chap. 3, regarding the detrimental effect of the bending on the collapse pressure of the pipes, but also show the dependency of the collapse on the load path.

Regarding the slotted pipes, the agreement between the finite element predictions and the laboratory observations is also very good. The detrimental effect of the slots on the structural resistance of the slotted liners is evident.

Finally, our analyses confirm that even when the applied axial tensile load matches the material yield load there still remains a non negligible capacity in the pipes for carrying external pressure.

References

1. Specification for Line Pipe (1992) API specification 5L, 40th edn.
2. Assanelli AP, Toscano RG, Johnson D, Dvorkin EN (2000) Experimental/numerical analysis of the collapse behavior of steel pipes. Eng Comput 17:459–486
3. Arbocz J, Babcock CD (1969) The effect of general imperfections on the buckling of cylindrical shell. ASME, J Appl Mechs 36:28–38
4. Arbocz J, Williams JG (1977) Imperfection surveys of a 10-ft diameter shell structure. AIAA J 15:949–956
5. Yeh MK, Kyriakides S (1988) Collapse of deepwater pipelines. ASME, J Energy Res Tech 110:1–11
6. Assanelli AP, López Turconi G (2001) Effect of measurements procedures on estimating geometrical parameters of pipes. In: 2001 Offshore technology conference, Paper OTC 13051, Houston, Texas
7. Toscano RG, Timms C, Dvorkin EN, DeGeer D (2003) Determination of the collapse and propagation pressure of ultra-deepwater pipelines. In: Proceedings of the OMAE 2003, 22nd international conference on offshore mechanics and artic engineering
8. Zimmerman T, DeGeer D (1999) Large pressure chamber tests ultra-deepwater pipe samples. Deepwater Technology, Supplement to World Oil August 67–72
9. DeGeer DD, Cheng JJ (2000) Predicting pipeline collapse resistance. In: Proceedings of the international pipeline conference, Calgary
10. Toscano RG, Amenta PM, Dvorkin EN (2002) Enhancement of the collapse resistance of tubular products for deepwater pipeline applications. In: IBC'S offshore pipeline technology, conference documentation
11. Kyriakides S (1994) Propagating instabilities in structures. Adv Appl Mech 30:67–189
12. Bathe K-J (1996) Finite element procedures. Prentice Hall, Upper Saddle River
13. Dvorkin EN, Bathe K-J (1984) A continuum mechanics based four-node shell element for general nonlinear analysis. Eng Comput 1:77–88
14. Bathe K-J, Dvorkin E (1985) A four-node plate bending element based on Mindlin/Reissner plate theory and a mixed interpolation. Int J Numer Methods Eng 21:367–383
15. Bathe K-J, Dvorkin EN (1986) A formulation of general shell elements—the use of mixed interpolation of tensorial components. Int J Numer Methods Eng 22:697–722
16. Bathe K-J, Dvorkin EN (1983) On the automatic solution of nonlinear finite element equations. Comput Struct 17:871–879
17. The ADINA SYSTEM (2012) Adina R&D. Watertown
18. Toscano RG, Gonzalez M, Dvorkin EN (2003) Validation of a finite element model that simulates the behavior of steel pipes under external pressure. J Pipeline Integr 2:74–84
19. Bulletin on Formulas and Calculations for Casing, Tubing, Drill Pipe, and Line Pipe Properties (1994) API bulletin 5C3 6th edn.

20. Det Norske Veritas (2000) Offshore standard OS-F101, Submarine pipeline systems
21. Suryanarayana S (s.f.). private communication
22. Toscano RG, Dvorkin EN (2011) Collapse of steel pipes under external pressure and axial tension. J Pipeline Eng 4:213–214

Chapter 5
Collapse and Post-Collapse Behavior of Deepwater Pipelines with Buckle Arrestors: Cross-Over Mechanisms

5.1 Introduction

Deepwater pipelines are normally subjected to external pressure and bending. They fail due to structural collapse when the external loading exceeds the pipes collapse limit surface. For seamless steel pipes, the influence on this limit surface of manufacturing imperfections has been thoroughly studied using finite element models that have been validated via laboratory full-scale tests, as it was shown in Chap. 4.

If by accident the collapse is initiated at a certain location, the collapse is either restrained to the collapse initiation section or it propagates along the pipeline, being this second alternative the most detrimental one for the pipeline integrity [1]. Since the external collapse propagation pressure is usually quite low in comparison with the external collapse pressure, it is necessary to build in the pipeline spaced reinforcements, usually steel rings, to act as arrestors for the collapse propagation.

Two different buckle arrestor cross-over mechanisms were identified in the literature: flattening and flipping. The occurrence of either cross-over mechanism is determined by the geometry of the pipes and of the arrestors [2].

In this chapter we focus on the analysis of the collapse and post-collapse behavior of pipelines reinforced with buckle arrestors; we develop finite element models to analyze the collapse, collapse propagation and cross-over pressures of reinforced pipes and we present an experimental validation of the models. In particular we consider the case of welded integral arrestors.

In Sect. 5.2 we describe the experimental facilities and the laboratory tests that we performed to determine, for different [*pipe-arrestor*] geometries, the collapse, propagation and cross-over pressures. In Sect. 5.3 we describe the finite element models that we developed to simulate the collapse tests and finally we compare the experimental results with the finite element ones, in order to validate the model.

The results we present in this chapter were published in [3].

E. N. Dvorkin and R. G. Toscano, *Finite Element Analysis of the Collapse and Post-Collapse Behavior of Steel Pipes: Applications to the Oil Industry*, SpringerBriefs in Computational Mechanics, DOI: 10.1007/978-3-642-37361-9_5,

5.2 Experimental Results

5.2.1 Experimental Set-up

The purpose of the laboratory tests developed for different combinations [*pipe* + *arrestor* + *pipe*] was to track the post-collapse equilibrium path for the assembly under external pressure and to determine from it the collapse and the cross-over pressures. For these tests we used the experimental set-up shown in Fig. 5.1.

Each sample had two pipes, one on each side of the arrestor; Fig. 5.2 shows the assembly of one of the samples. For each side, an L/OD ratio greater than 7.5 was used in order to minimize the end—effects on the collapse loads.

Two solid end-caps were welded on each end. The internal section of the end-caps was shaped to avoid localized failure during propagation. The shape of this section was derived from the finite element results of a free propagating buckle (Fig. 5.12).

Figure 5.3 shows a detailed drawing and a photograph of the collapse chamber.

Each specimen was completely filled with water before the beginning of the test. From a 25 mm hole in one of the end-caps (Fig. 5.4) the displaced water was directed to a container connected to a load cell. The load variation in the load cell is proportional to the displaced water and therefore to the variation of the specimen inner volume.

To localize the buckle initiation, we milled a groove on one of the pipes (upstream pipe) as shown in Fig. 5.5.

In Fig. 5.6 we present a detail of the geometry of the arrestor and we define the dimensions and steel grade of the four tested samples.

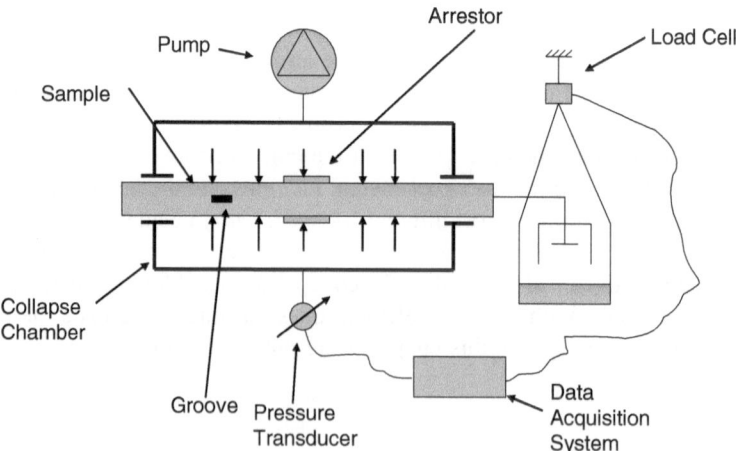

Fig. 5.1 Experimental set-up

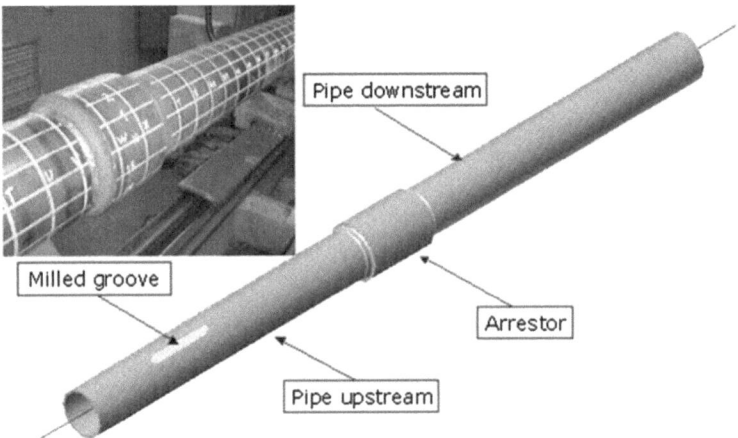

Fig. 5.2 Pipes and arrestor assembly

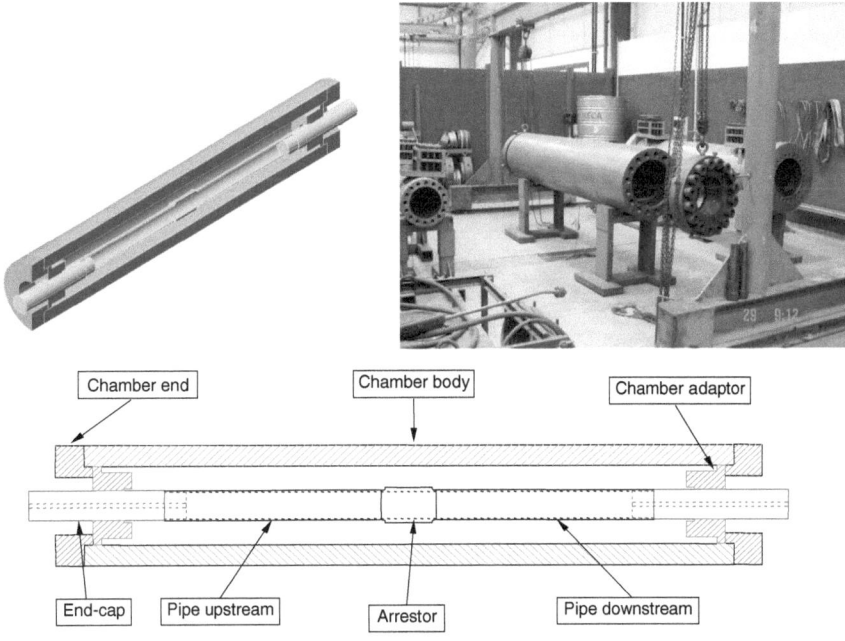

Fig. 5.3 Collapse chamber

The steel Grade 6 defined by the standard ASTM A-333 has a minimum yield stress of 240 MPa and a minimum ultimate stress of 414 MPa.

The steel Grade X42 defined by the standard API-5L has a minimum yield stress of 290 MPa and a minimum ultimate stress of 414 MPa.

Fig. 5.4 End-caps welded to the pipe

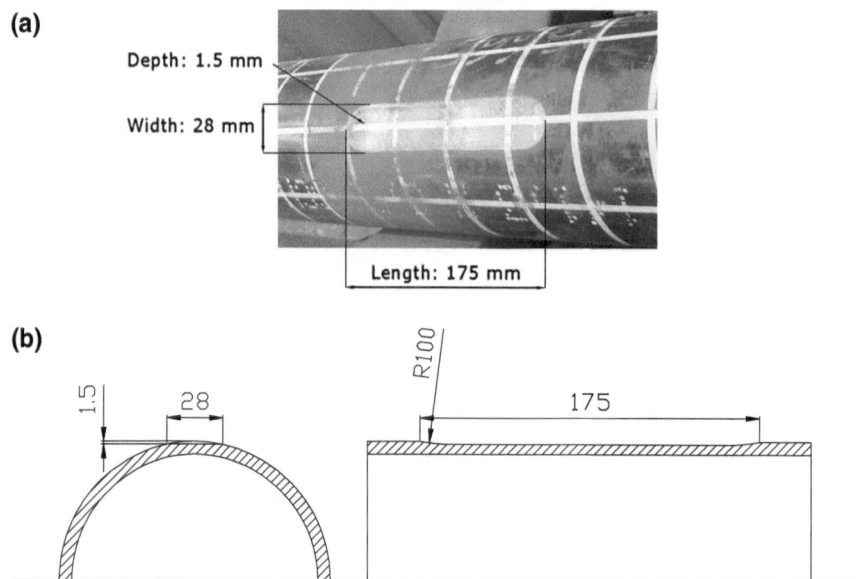

Fig. 5.5 Groove machined on the upstream pipe to localize the collapse initiation. **a** General view. **b** Detailed sections

During the tests, we continuously increased the external pressure as in a standard collapse test; after the collapse the pumping continued through the upstream propagation, cross-over of the arrestor and downstream propagation.

5.2.2 Geometrical Characterization of the Tested Samples

The outer surface of the samples was mapped using the shapemeter [4]. The OD Fourier mode decompositions and the thickness maps of the sample 1 are shown in the Appendix.

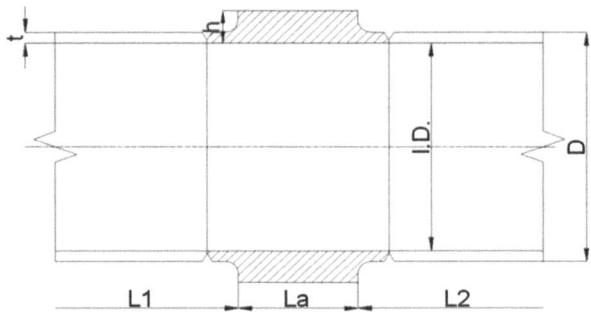

Sample	Pipe OD [mm]	Pipe thickness (t) [mm]	Pipe steel grade	Arrestor (h/t)	Arrestor (La/D)	Arrestor steel grade	Sample length [mm]	Expected cross-over mechanism
1	141.3	6.55	X42	2.0	0.50	6 (ASTM A-333)	2250	Flattening
2	141.3	6.55	X42	2.5	0.50	X42	2250	Flattening
3	141.3	6.55	X42	3.0	0.75	X42	2274	Flipping
4	141.3	6.55	X42	3.0	1.00	X42	2330	Flipping

Fig. 5.6 Welded arrestors geometry and materials

5.2.3 Mechanical Characterization of the Tested Samples

For all the pipe and arrestor materials we determined,

- stress–strain curves (longitudinal tensile tests since the thickness of the pipes was too small for hoop samples),
- hoop residual stresses (evaluated using slit ring tests).

In Table 5.1 we summarize the residual stress values.

5.3 The Finite Element Model

As in the models described previously, we simulated the external pressure collapse test using the MITC4 [5–7] shell element implemented for finite elasto-plastic strains in the ADINA system [8]. The numerical model was developed using a material and geometrical nonlinear formulation, which takes into account large

	Sample	$\dfrac{Measured.max.Residual.Stresses}{Measured.Yield.Stress}$
Table 5.1 Residual stresses measured using the slit ring test	1	0.39
	2	0.47
	3	0.47
	4	0.49

displacements/rotations and finite strains, since it was discussed in [9] that even though the strains during post-collapse regime are rather small, at concentrated locations they can attain quite large values, as shown in Fig. 5.7.

In previous analyses we observed that when using an infinitesimal strains formulation we get results that have an excellent match with the experimental determinations; to confirm this assessment, in this chapter we compare the experimental results with the numerical results obtained under the assumption of infinitesimal strains and under the assumption of finite strains [10].

The model incorporates the following features [3],

- von Mises elasto-plastic material model with isotropic multi-linear hardening. In Figs. 5.8 and 5.9 we show, for one of the tested samples, the experimental stress–strain curves and its fitting using a multilinear hardening model,
- contact elements on the pipe inner surface in order to prevent its inter-penetration in the post-collapse and propagation regimes,
- nonlinear equilibrium path tracing via the algorithm developed in [11],
- hoop residual stresses modeled with the technique discussed in Chap. 3.

Fig. 5.7 Typical post-collapse Hencky strains distribution

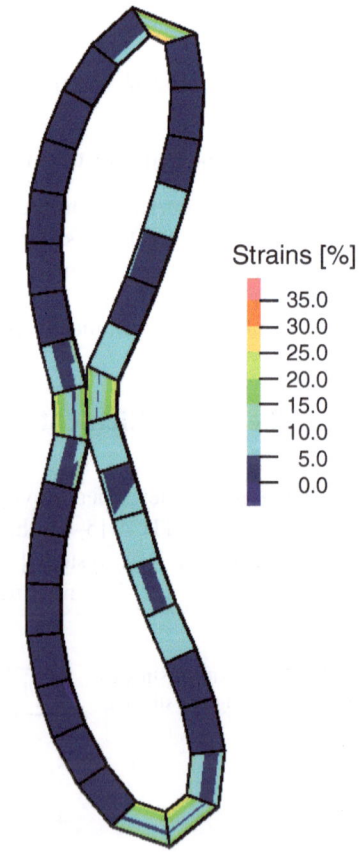

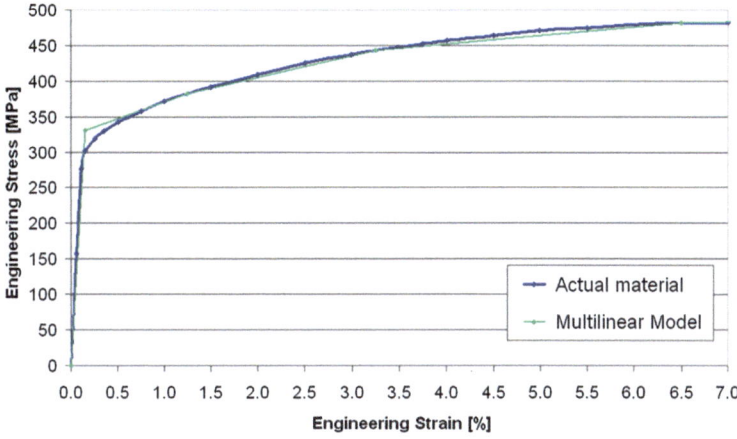

Fig. 5.8 Material model for the pipe segments in sample no 1. Experimental and interpolated curves

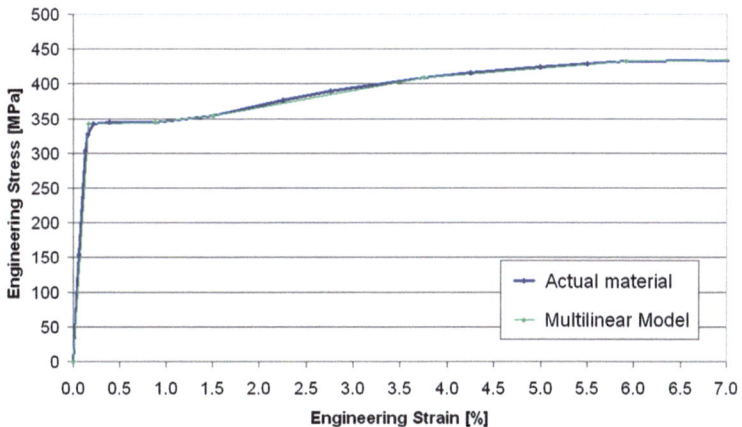

Fig. 5.9 Material model for the arrestor in sample no 1. Experimental and interpolated curves

In Fig. 5.10 we present the finite element mesh, 8500 elements and 42,500 d.o.f, a detail of the mesh in the [*pipes-arrestor*] transition which was modeled using variable thickness elements [12] and a detail of the end-caps modeling; there are contact elements between the end-caps and the pipes.

Figure 5.11 shows the contact pressure distribution in the third sample, immediately after the cross-over; this figure includes a detail of the plug cone.

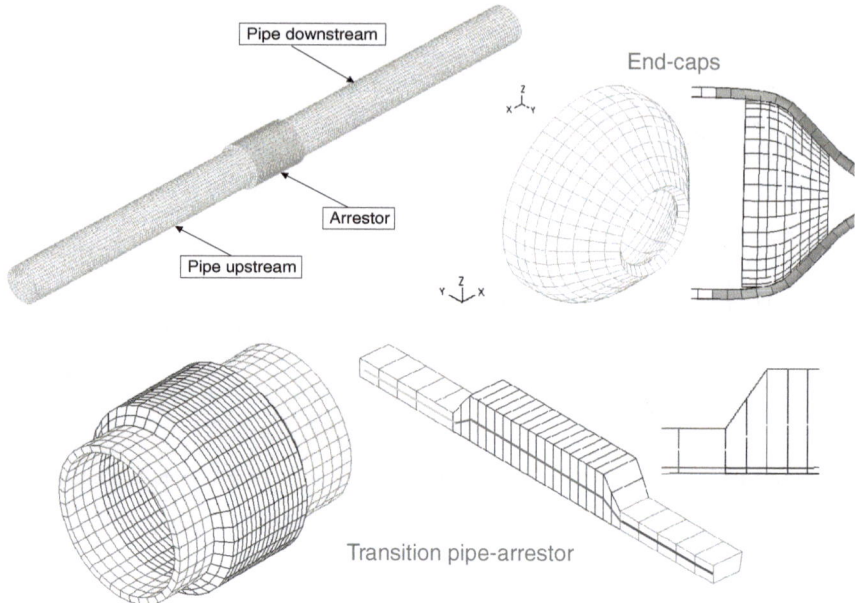

Fig. 5.10 Pipe-arrestor finite element mesh

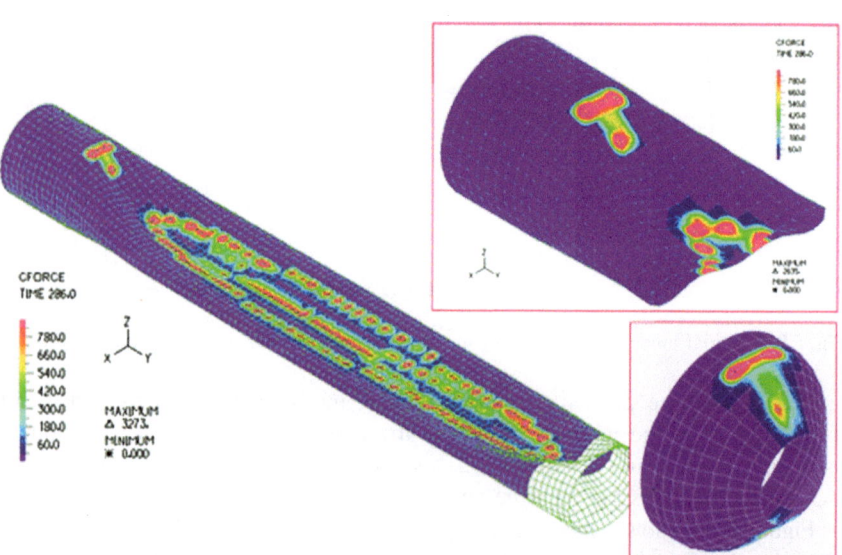

Fig. 5.11 Sample 3. Contact forces for the upstream pipe and plug zone

5.4 Validation of the Finite Element Results

In this section we discuss the validation of our finite element results by comparing them with experimental determinations that we obtained using the set-up described above.

5.4.1 Exploring the Finite Element Model

In order to explore the behavior of our finite element model, first we analyze two perfect samples, without residual stresses. In the first one we expect the collapse buckle to cross the arrestor with a flattening mode while in the second case, because of the higher stiffness of the arrestor, we expect a flipping mode. That supposition is based in previous parametric analyses, where we considered different ratios between the pipes versus the arrestor stiffness.

In each case we consider an imperfection, centered at a distance of 236.1 mm from the upstream pipe end, with a shape [13],

$$w_0 = -\Delta_0 \left(\frac{OD}{2}\right) \exp\left[-\beta\left(\frac{x}{OD}\right)^2\right] \cos(2\theta)$$

where,

w_0 radial displacement,
θ polar angle,
Δ_0 imperfection amplitude parameter (0.002 mm),
β parameter that decides the extent of the imperfection, in our case (2.32 OD),
OD outside diameter,
x axial coordinate.

In Fig. 5.12 we present the finite element predicted deformed shapes for a [pipes-arrestor] system exhibiting the flattening cross-over mechanism and in Fig. 5.13 we show the predicted deformed shapes for a system presenting the flipping cross-over mechanism.

In both cases we plot the external pressure as a function of the internal volume variation.

Considering the [External pressure-volume variation] diagrams predicted by the finite element models, in each case we observe,

- the test starts at point "1" and while the external pressure grows the sample maintains its original shape and therefore there is a very small internal volume variation. Then the point of maximum pressure is reached ("collapse pressure") and the sample rapidly changes its cross-section shape; while the collapse buckle grows in its amplitude and extension in the upstream pipe axial direction,

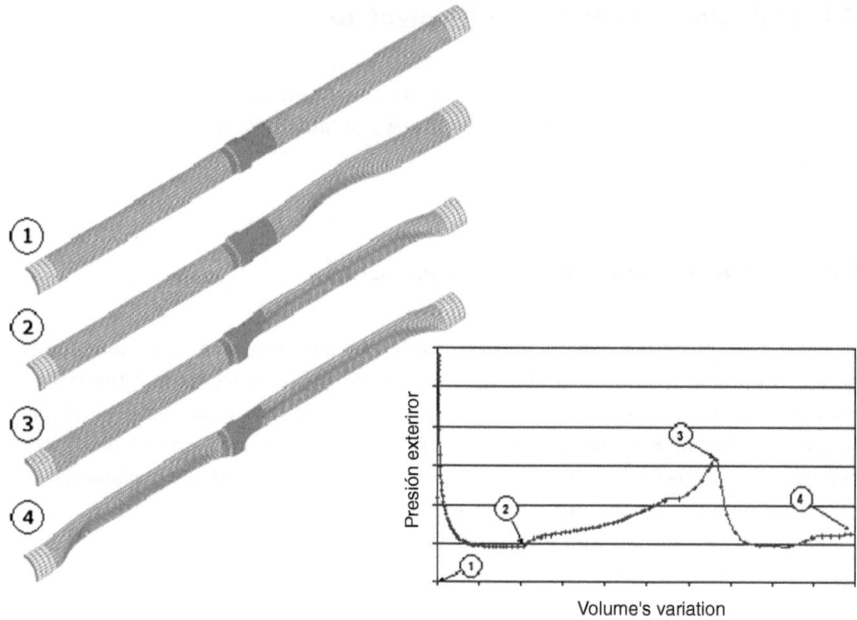

Fig. 5.12 Finite element results for the case presenting a flattening cross-over

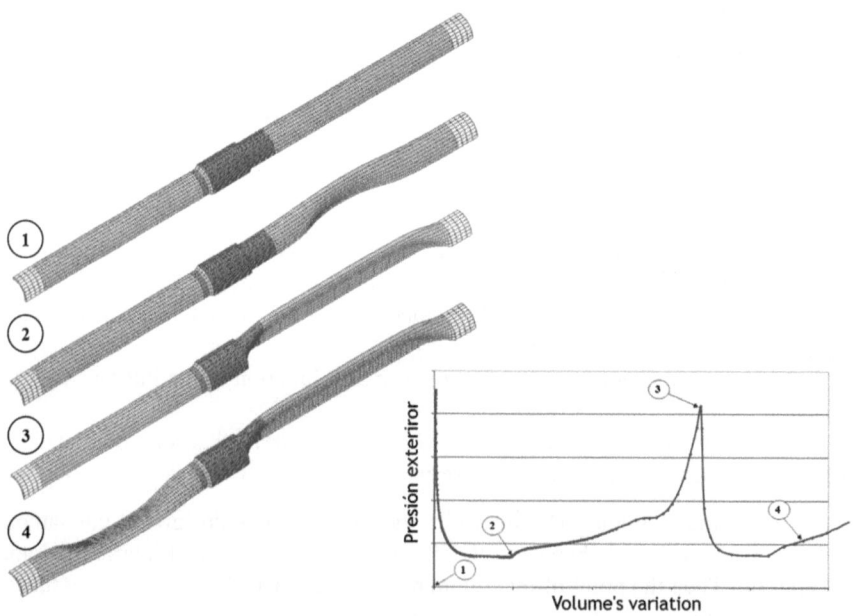

Fig. 5.13 Finite element results for the case presenting a flipping cross-over

the external equilibrium pressure drops. At some point the collapse buckle extension starts to grow under constant external pressure [1],

- at "2" opposite points located on the inner surface of the upstream pipe establish contact and afterwards, while the contact area extends, the external equilibrium pressure increases,
- while the collapse buckle in the upstream pipe approaches the arrestor the external equilibrium pressure keeps increasing but the downstream pipe does not collapse,
- at point "3" ("cross-over pressure") the collapse buckle crosses the arrestor and the downstream pipe collapses,
- afterwards, the collapse buckle propagates through the downstream pipe.

It is important to notice that in the case with the flattening cross-over mechanism the upstream and downstream pipe have their collapsed sections with the same orientation while in the case with the flipping cross-over mechanism the collapse sections form an angle close to ninety degrees. It is also important to notice that the relation [*cross-over pressure/collapse pressure*] is much higher for the flipping case than for the flattening case.

5.4.2 Comparison Between the Finite Element and Experimental Results

The four samples tested in the laboratory were modeled and the [*external pressure-volume variation*] equilibrium paths were determined.

In Table 5.2 we compare the FEM and experimental results. The FEM analyses were performed under the assumption of both, finite strain and infinitesimal strains.

It is important to point out that the finite element results indicated in this table were obtained considering that the residual stresses in the two pipe sections are the residual stresses measured in the full length pipe. The modifications in residual stresses induced by the pipe cutting, the welding and groove machining were not introduced in the model. The effect of the residual stresses on the collapse pressure was discussed in previous publications [4, 9, 10, 14, 15] as well as in Chap. 3. While this effect is quite important, we found with our numerical experimentation,

Table 5.2 Validation of the numerical model (finite strain assumption)

Sample	Collapse_pressure : $\frac{FEM_{finite\_strains}}{LAB}$	Cross_over_pressure : $\frac{FEM_{finite\_strains}}{LAB}$
1	0.924	1.004
2	0.928	0.985
3	0.951	0.926
4	0.852	0.883

that the effect of the residual stresses on the cross-over pressure is not so relevant, as shown by the model results that we present in Fig. 5.14.

In Figs. 5.15 and 5.16 we compare, for samples 1 and 2 (flattening), the experimentally determined and FEM predicted equilibrium paths under the assumptions of finite strains and infinitesimal strains,

- during the laboratory determination for the first sample some water was spilled out of the measurement system, a fact that explains the shift observed, in the horizontal axis, between the FEM and experimental results,
- for the second sample the agreement between the FEM and experimental results is very good,

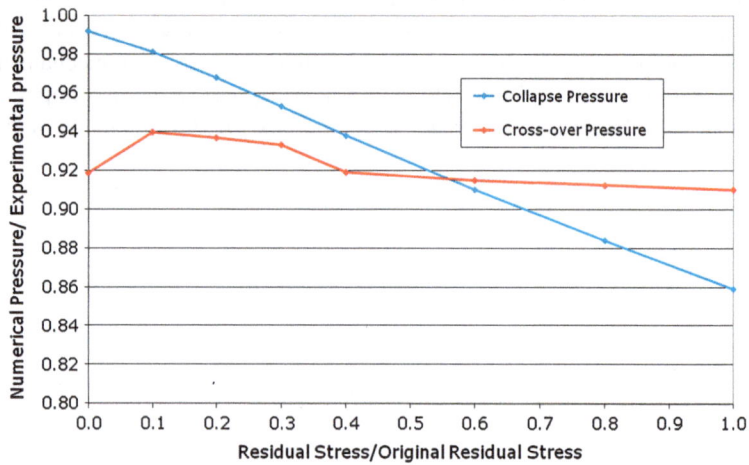

Fig. 5.14 Sample 4. Residual stresses effect on the collapse and cross-over pressures

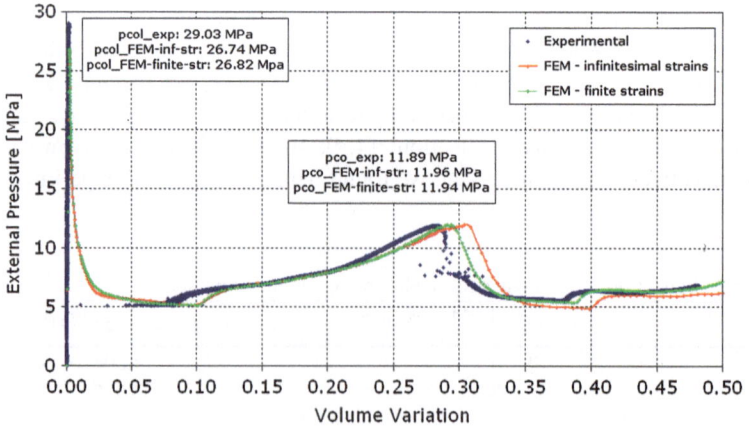

Fig. 5.15 Sample 1. FEM versus experimental results (flattening cross-over)

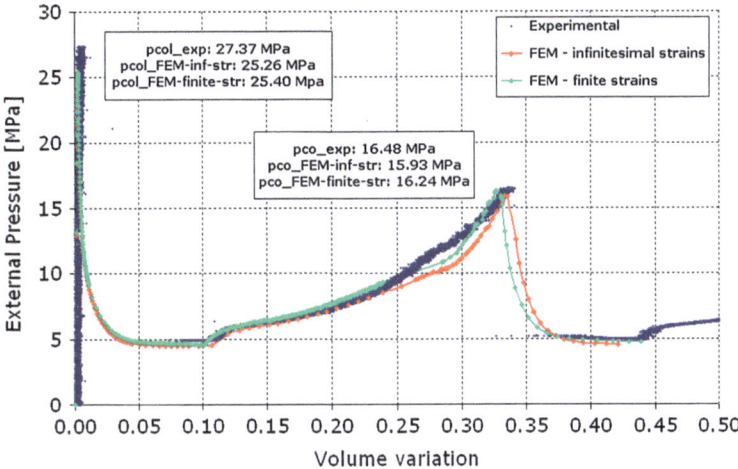

Fig. 5.16 Sample 2. FEM versus experimental results (flattening cross-over)

Fig. 5.17 Sample 2.
Deformed mesh (flattening
cross-over). Equivalent
plastic strains

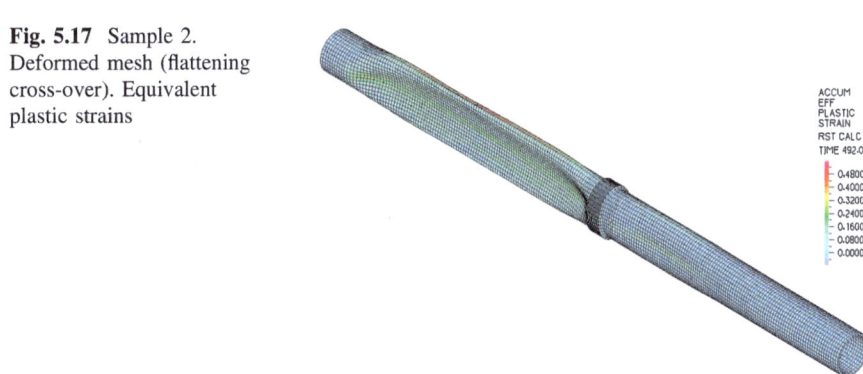

- the results obtained using FEM under the assumptions of finite and infinitesimal strains are very close.

In Fig. 5.17 we present, for sample 2, the deformed finite element mesh after cross-over.

In Figs. 5.18 and 5.19 we present the same comparison for samples 3 and 4 (flipping). Again, the agreement between FEM and experimental results is very good and again the results obtained using FEM under the assumptions of finite and infinitesimal strains are very close.

In Fig. 5.20 we present, for sample 4, the deformed finite element mesh after cross-over.

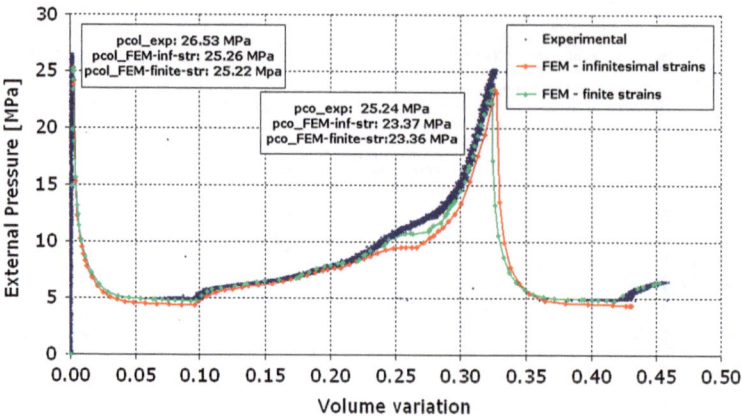

Fig. 5.18 Sample 3. FEM versus experimental results (flipping cross-over)

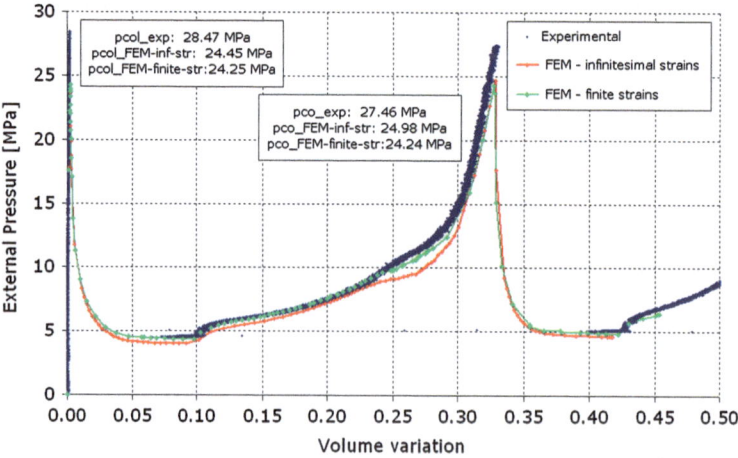

Fig. 5.19 Sample 4. FEM vs. experimental results (flipping cross-over)

Fig. 5.20 Sample 4. Deformed mesh (flipping cross-over). Equivalent plastic strains

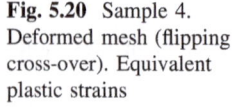
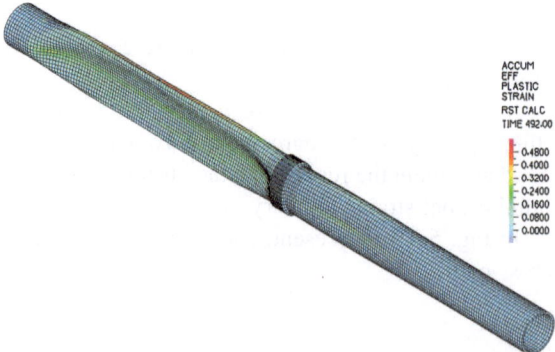

Fig. 5.21 Sample 2.
Experimentally observed and
FEM predicted shapes of
collapsed pipes after a
flattening cross-over

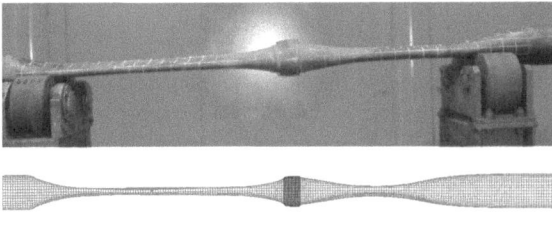

Fig. 5.22 Sample 3.
Experimentally observed and
FEM predicted shapes of
collapsed pipes after a
flipping cross-over

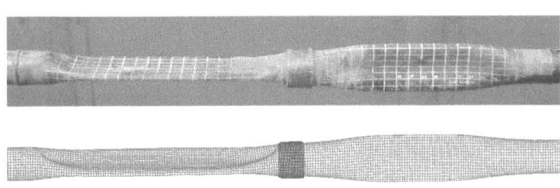

Finally, in Fig. 5.21 we compare the experimentally observed and FEM predicted shapes for a case where the cross-over mechanism was flattening. In Fig. 5.22 we make the same comparison for a case in which the cross-over mechanism was flipping. In both cases the agreement between numerical and experimental results is excellent.

It is interesting to notice that in samples 2 and 4 the plastic strains in the deformed section knee are very high; in our case the elements were removed when the equivalent plastic strain reaches 100 %; however more sophisticated criteria for the material damage can be implemented [16].

5.5 Main Observations

The two collapse modes reported in the literature, the flattening and the flipping modes, were correctly identified in our simulations.

The agreement between the finite element predictions and the laboratory observations, both for the collapse and cross-over pressure, is very good; hence, finite element models can be used as a reliable engineering tool to assess the performance of integral ring buckle arrestors for steel pipes.

Regarding the use of a finite strain elasto-plastic formulation or of infinitesimal strains, our numerical results consistently show that if the analysis purpose is the determination of the collapse and cross-over pressures, the use of the less expensive infinitesimal strain model is amply justified; however, if detailed information on the strain/stresses in the collapse buckle is sought, the more resources consuming finite strain models should be used.

References

1. Palmer AC, Martin JH (1975) Buckle propagation in submarine pipelines. Nature 254:46–48
2. Park TD, Kyriakides S (1997) On the performance of integral buckle arrestors for offshore pipelines. Int J Mech Sci 39:643–669
3. Toscano RG, Mantovano LO, Amenta PM, Charreau R, Johnson D, Assanelli AP, Dvorkin EN (2008) Collapse arrestors for deepwater pipelines. Cross-over mechanisms. Comput Struct 86:728–743
4. Assanelli AP, Toscano RG, Johnson D, Dvorkin EN (2001) Experimental/numerical analysis of the collapse behavior of steel pipes. Eng Comput 17:459–486
5. Dvorkin EN, Bathe K-J (1984) A continuum mechanics based four-node shell element for general nonlinear analysis. Eng Comput 1:77–88
6. Bathe K-J, Dvorkin EN (1985) A four-node plate bending element based on Mindlin/Reissner plate theory and a mixed interpolation. Int J Numer Methods Eng 21:367–383
7. Bathe K-J, Dvorkin EN (1986) A formulation of general shell elements—the use of mixed interpolation of tensorial components. Int J Numer Methods Eng 22:697–722
8. The ADINA system. ADINA R&D, Watertown
9. Toscano RG, Gonzalez M, Dvorkin EN (2003) Validation of a finite element model that simulates the behavior of steel pipes under external pressure. J Pipeline Integr 2:74–78
10. Toscano RG, Mantovano LO, Dvorkin EN (2004) On the numerical calculation of collapse and collapse propagation pressure of steel deep-water pipelines under external pressure and bending. Experimental verification of the finite element results. In: Proceedings 4th international conference on pipeline technology, Ostend, Belgium, pp 1417–1428
11. Bathe K-J, Dvorkin EN (1983) On the automatic solution of nonlinear finite element equations. Comput Struct 17:871–879
12. Bathe K-J (1996) Finite element procedures. Prentice Hall, Upper Saddle River
13. Netto TA, Kyriakides S (2000) Dynamic performance of integral buckle arrestors for offshore pipelines. Part II: Analysis. Int J Mech Sci 42:1425–1452
14. Toscano RG, Amenta PM, Dvorkin EN (2002) Enhancement of the collapse resistance of tubular products for deepwater pipeline applications. In: IBC'S Offshore Pipeline Technology, conference documentation
15. Toscano RG, Timms C, Dvorkin EN, DeGeer D (2003) Determination of the collapse and propagation pressure of ultra-deepwater pipelines. OMAE 2003, 22nd international conference on offshore mechanics and Artic engineering
16. Bao Y, Wierzbicki T (2004) A comparative study on various ductile crack formation criteria. J Eng Mater Technol ASME 126:314–324

Chapter 6
Conclusions

Nowadays important technological decisions are based on the results of computational models that simulate the collapse and post-collapse behavior of steel pipes. Since these technological decisions have a large influence on the ecological impact of pipeline installations and other industrial facilities, on labor conditions and on revenues, it is important that those computational models are highly reliable.

Therefore, it is of the utmost importance that sound computational mechanics formulations are used with adequate material data; also it is necessary that the computational results are subjected to experimental validation.

In this book we proposed some guidelines for the development of finite element models that provide the link between manufacturing tolerances and performance predictions.

Some of the aspects that we discussed are the usage of 2D and 3D models, the required model nonlinearities, the usage of follower type loads, the material modeling, the modeling of residual stresses and the code verification and model validation.

6.1 The Usage of 2D and 3D Models

Material properties, residual stresses and pipe dimensions like eccentricity, out-of roundness, thickness, etc. vary along the length of a given pipe.

When the collapse behavior of a specific pipe is investigated, a 3D model that incorporates a detailed geometrical and material description needs to be developed.

However, we may also need to perform parametric studies to investigate, for example, the effect of manufacturing tolerances on the collapse and propagation pressures; in these cases we consider an infinite pipe with uniform properties along its length and we use a 2D plane strain model built using continuum elements or 3D short shell models, if we need to include bending in the analysis.

In Sect. 3.2 we described the model we developed using the 2D continuum elements QMITC. This model was very useful to study the effect of ovality,

E. N. Dvorkin and R. G. Toscano, *Finite Element Analysis of the Collapse and Post-Collapse Behavior of Steel Pipes: Applications to the Oil Industry*, SpringerBriefs in Computational Mechanics, DOI: 10.1007/978-3-642-37361-9_6,

eccentricity and residual stresses on the external collapse pressure. To assess on the quality of the mesh we analyzed the plane strain collapse of an infinite pipe and we compared our numerical results with the analytical results obtained using theoretical formulas. The comparison showed that the proposed 2D mesh of QMITC elements is accurate enough to represent the collapse of very long specimens.

The ovality is considered to be concentrated in the shape corresponding to the first elastic buckling mode and the eccentricity is modeled considering non-coincident OD and ID centers.

For the range of (D/t) values that are relevant for pipelines, we can model the collapse and post-collapse behavior of the pipes using shell elements. In particular we selected a shell element that is free from the locking problem: the MITC4 element. In the code ADINA this element was implemented improving its in-surface behavior incorporating incompatible modes.

In Sect. 3.3, to include bending in the analyses, we developed a numerical model to simulate the behavior of a very long tube (infinite tube) and determine its pre and post-collapse equilibrium path. Using this model we performed parametric studies in order to investigate the significance of the different geometrical imperfections and of the residual stresses on the collapse and collapse propagation pressures.

The 3D finite element models of finite pipes were developed to overcome the limitations of the simpler models described above.

It is important to take into account that when the sample is long enough (L/D > 10) the end conditions have only a very small influence on the collapse pressure.

6.2 Nonlinearities

It is also necessary to decide on the nonlinearities that we must include in our finite element models to be able to predict the collapse of steel pipes under external pressure and to track their post-collapse behavior.

Since we need to predict collapse, we have to use a geometrically nonlinear analysis considering large displacements/rotations, that is to say we have to fulfill the equilibrium equations in the deformed configuration. As it was demonstrated in Sect. 5.4, even if very high strains are developed at localized points, the general behavior of the post-collapse response can be determined without resorting to the more expensive finite strain models.

In the range of (OD/t) values that are within our scope, the collapse is an elastic–plastic collapse, that is to say plasticity is developed before and after collapse; hence the material nonlinearity has to be included in the analysis.

To track the collapse and post-collapse response of the pipes we use an algorithm that iterates in the load–displacement space.

6.3 Follower Loads

It is important to consider follower loads to model the effect of the external hydrostatic pressure, since the consideration of fixed-direction loads results in important errors when predicting collapse pressures.

6.4 Material Modeling

In our models we use von Mises associated elasto-plastic material models with isotropic hardening. We model the hardening using bilinear or multi-lineal models. Even though it is clear that more sophisticated hardening models can be used, this very simple model has been very successful in the prediction of collapse and post-collapse pipe behaviors, as it was shown in Chaps. 4 and 5.

6.5 Modeling of Residual Stresses

In our analyses we considered a linear residual stresses distribution across the pipe wall thickness. In Sect. 3.4 we checked the modeling of the residual stresses distribution by modeling a slit ring test using the ADINA [1] "element birth and death" feature.

6.6 Code Verification and Model Validation

In the verification process we have to prove that we are solving the equations right, and therefore this is a mathematical step [2]. In this step we have to show that our numerical scheme is convergent and stable.

It is important to notice that the verification process is not only related to a numerical procedure but also to its actual implementation in software (either commercial software or an in-house one) [2]. For the verification of the numerical formulations and algorithms we relied on the verification of the software developers [3].

In the validation process we have to prove that we are solving the right equations, and therefore it is an engineering step [2].

We do validate neither a formulation nor software: we validate the usage of verified software by an analyst in the simulation of a specific process. We have to validate the complete procedure.

In Chap. 4 we developed a complete validation of the collapse and post-collapse analyses of pipes under external pressure plus bending while in Chap. 5

we validated the analysis of pipes with collapse arrestors under external pressure only. Regarding the latter case, the numerical results not only match the experimental values of collapse pressure, collapse propagation pressure and cross-over pressure but also the cross-over collapse modes, flattening and flipping.

Summarizing, a methodology for using the finite element method as a robust engineering tool for analyzing the effect of the manufacturing tolerance on the collapse and post-collapse behavior of steel pipes was discussed and illustrated with practical examples.

References

1. The ADINA SYSTEM (2012) Adina R&D, Watertown
2. Roache PJ (1998) Verification and validation in computational science and engineering. Hermosa Publishers, Hermosa
3. ADINA Verification Manual (2010) ADINA R&S, INC. Watertown

Appendix
Imperfections Measuring System

A.1 Introduction

The Imperfections Measuring System (IMS) or "*shapemeter*" is based on publications by Arbocz and co-workers [1–3] and we developed it to validate our finite element models comparing numerical with experimental results [4]. The IMS was used to survey the geometry of the pipes that were tested in collapse facilities at Tenaris Siderca (Argentina) and at C-FER Technologies (Canada). A photograph of the IMS is shown in Fig. A.1.

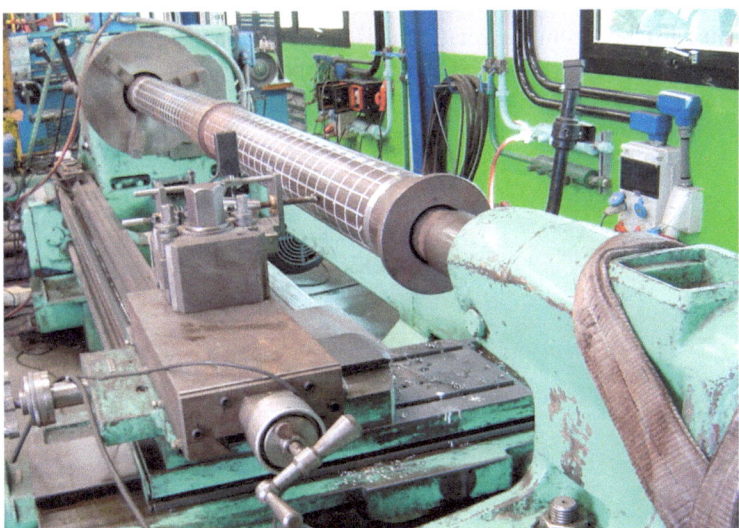

Fig. A.1 The IMS (or shapemeter)

E. N. Dvorkin and R. G. Toscano, *Finite Element Analysis of the Collapse and Post-Collapse Behavior of Steel Pipes: Applications to the Oil Industry*, SpringerBriefs in Computational Mechanics, DOI: 10.1007/978-3-642-37361-9, © The Author(s) 2013

A.2 Mapping of the Sample External Surfaces

The samples are rotated in the lathe and on its carriage an linear variable displacement transducer (LVDT) is placed. The LVDT touches the external surface of the rotating samples and at regular intervals of time its radial position is recorded; also, the angular position of the samples is recorded at the same time intervals by a rotary encoder.

We developed an algorithm to obtain from the acquired data a Fourier series description of the external surface of the samples [4].

A.2.1 Algorithm to Process the Data Acquired with the IMS

The data are acquired along a spiral path; however, in subsequent analyses we will consider that the points corresponding to a turn are located on a planar section, at an axial distance z_k from an arbitrary origin. As the pitch of the spiral is less than half of the typical wall thickness under analysis, this assumption is valid for the purpose of modeling the collapse tests.

The data are fitted to a perfect circle (of unknown center and radius) through a least squares method [3]. This approach is consistent with the subsequent Fourier decomposition (see Fig. A.2).

Input data

r_j radial distance from the rotation axis to the external surface, jth data point.

q_j total turns corresponding to the jth data point, measured from an arbitrary defined zero.

Fig. A.2 Algorithm to process the data acquired with the LVDT

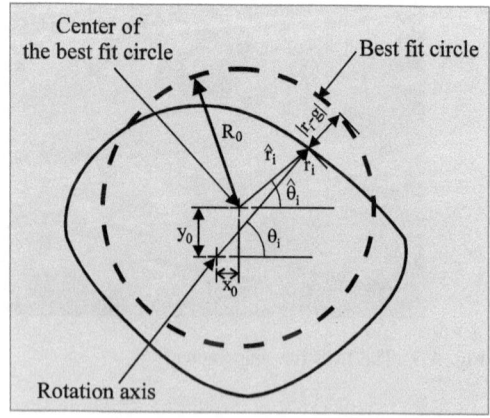

Algorithm
Initial data reduction

We can define $k = int(q_j)$ the kth turn. For this turn we have,

$$z_k = \Delta z \cdot int(q_j)$$
$$\theta_i^k = 2\pi(q_j - int(q_j))$$
$$r_i^k = r_j$$

where $i = 1$ for the first j which satisfies $(q - int(q_j)) > 0$ (indication of a new turn). Δz indicates the axial advance per turn. The number of data points per turn is not constant.

Fit to best circle

For the kth section we can define a "best-fit circle", with radius R_0 and with its center located at (x_0, y_0) in a Cartesian system, contained in the section plane and with its origin at the section rotation center [3]. The superindex k in θ_i^k and r_i^k is omitted.

For determining R_0, x_0 and y_0 we solve the following minimization problem,

$$(R_0, x_0, y_0) = \arg[\min E_2(R_0, x_0, y_0)]$$
$$E_2 = \sum_i [r_i - g(\theta_i, R_0, x_0, y_0)]^2$$

$$g(\theta_i, R_0, x_0, y_0) = (x_0 \cos \theta_i + y_0 \sin \theta_i) + \sqrt{R_0^2 - (x_0 \sin \theta_i - y_0 \cos \theta_i)^2}$$

To solve the above nonlinear minimization problem we apply the Levemberg-Marquard method [5], using as first trial a simplified (linearized) solution in which the expression for g reduces to [6],

$$g_{lin}(\theta_i, R_0, x_0, y_0) = (x_0 \cos \theta_i + y_0 \sin \theta_i) + R_0$$

Data reduction to new center

Once the center of the "best-fit circle" is determined we reduce the acquired data to it,

$$\hat{x}_i = r_i \cos \theta_i - x_0$$
$$\hat{y}_i = r_i \sin \theta_i - y_0$$
$$\hat{r}_i = \sqrt{\hat{x}_i^2 + \hat{y}_i^2}$$
$$\hat{\theta}_i = \tan^{-1}\left(\frac{\hat{y}_i}{\hat{x}_i}\right)$$

Fourier transform

We expand using a discrete Fourier transform,

$$\hat{a}_j = \frac{1}{\pi} \sum_{k=1}^{M} \left[\hat{r}_k \cos\left(j\hat{\theta}_k\right) \Delta \hat{\theta}_k\right]$$

$$\hat{b}_j = \frac{1}{\pi} \sum_{k=1}^{M} \left[\hat{r}_k \sin\left(j\hat{\theta}_k\right) \Delta \hat{\theta}_k\right]$$

where M is the number of samples taken in each turn (360 on average).

Shape reconstruction

$$\hat{r}(\theta) = R_0 + \sum_{j=1}^{N} \left[\hat{a}_j \cos(j\theta) + \hat{b}_j \sin(j\theta) \right]$$

where N is the number of modes used in the reconstruction of the shape.

Sampling theorems [7] put a limit on the maximum value of N that can be used (in our case N < (M/2) $\simeq$ 180, typically 50). For practical purposes we define the amplitude of the j mode of the Fourier decomposition as,

$$A_j = \sqrt{\hat{a}_j^2 + \hat{b}_j^2}.$$

A.3 Mapping of the Sample Wall Thickness

A regular mesh is drawn on the samples external surface (see Fig. A.1) and at the mesh nodes the thickness is measured using standard manual ultrasonic gages.

A.4 Deepwater Pipelines: Measurements

The scope of this Section is to describe the results of the geometrical survey of the 9 samples of the tests described in Chap. 4. The geometrical survey is composed by the mapping of the external surface and the measurement of the wall thicknesses.

The results of the topography mapping of the external surface, the modal analysis of the circular deviations and the wall thickness are analyzed. The samples belong to two different mills.

Table A.1 summarizes the main characteristics of the samples.

Pipe Survey
Fourier analysis of the circular deviations
Figure A.3 shows, for samples 1–3, the Fourier series decomposition of the deviations with respect to the average diameter.

Table A.1 Samples description

Sample	Nominal OD [mm]	Nominal thickness [mm]
1	353.00	22
2	353.00	22
3	353.00	22
4	323.85	17.65
5	323.85	17.65
6	323.85	17.65
7	323.85	20.30
8	323.85	20.30
9	323.85	20.30

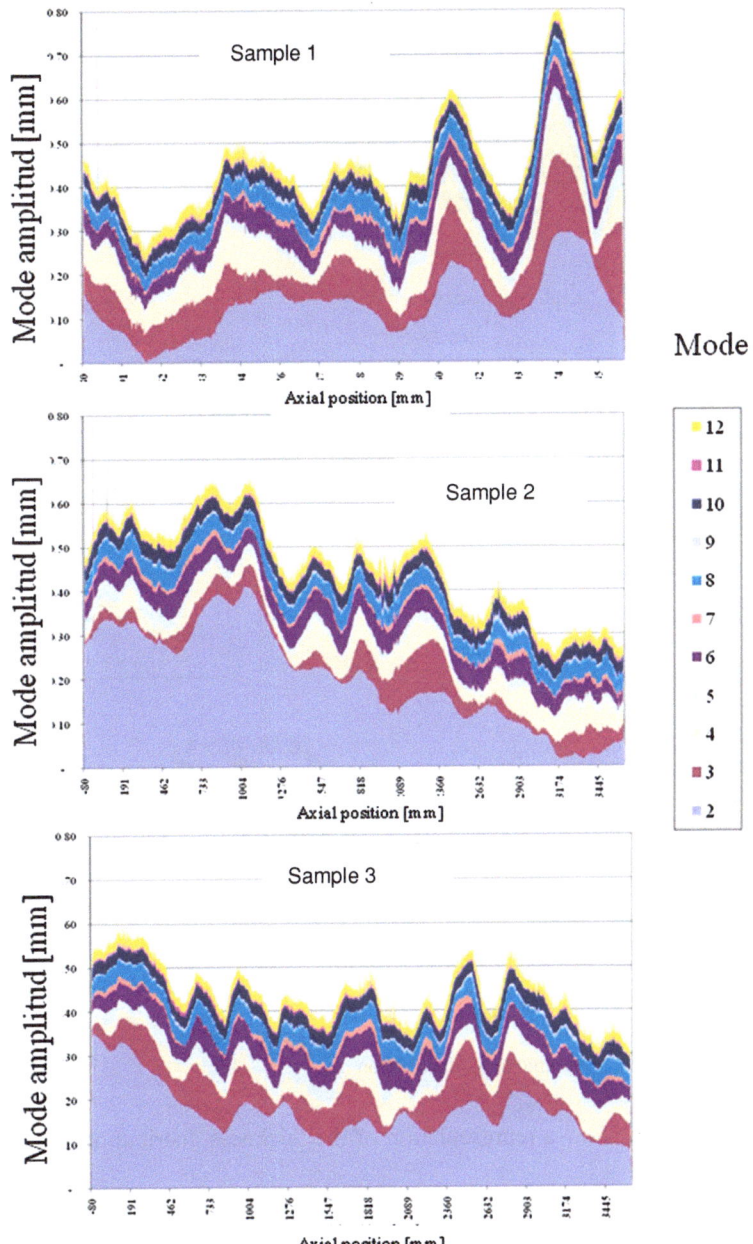

Fig. A.3 Samples 1–3: modal analysis: mode amplitude distribution along the sample

The specimens are 4 m long. The grid has 48 longitudinal sections with a 75 mm spacing, and 16 circumferential generatrices, giving 768 grid points.

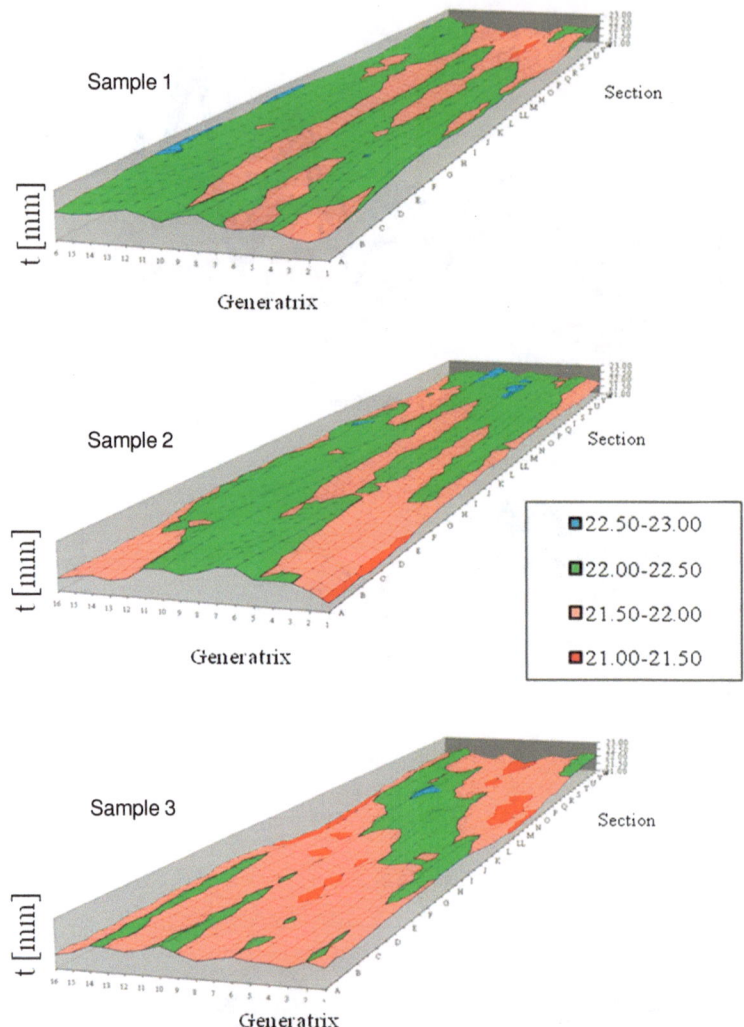

Fig. A.4 Samples 1–3, wall thickness distribution

In Fig. A.4 we show a representation of the thickness distribution for the first three samples.

A.5 Deepwater Pipelines with Buckle Arrestors: Measurements

For a sample made of two pipes with a welded intermediate arrestor the outside surface Fourier decomposition is shown in Fig. A.5 and the thickness distribution of the two pipes in Fig. A.6.

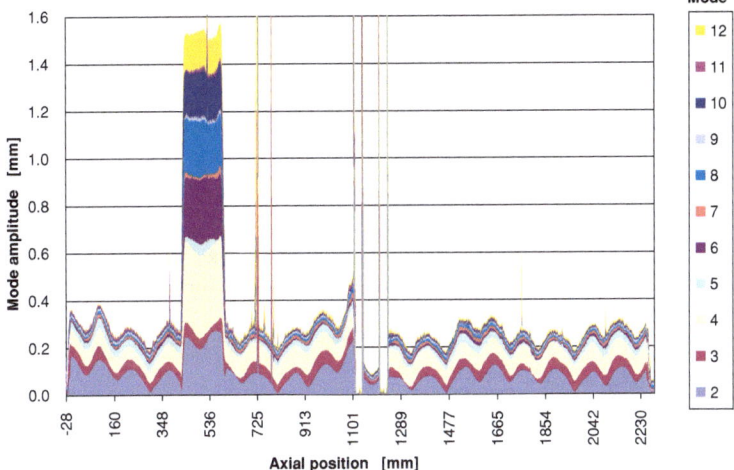

Fig. A.5 Outside surface Fourier decomposition of sample #1

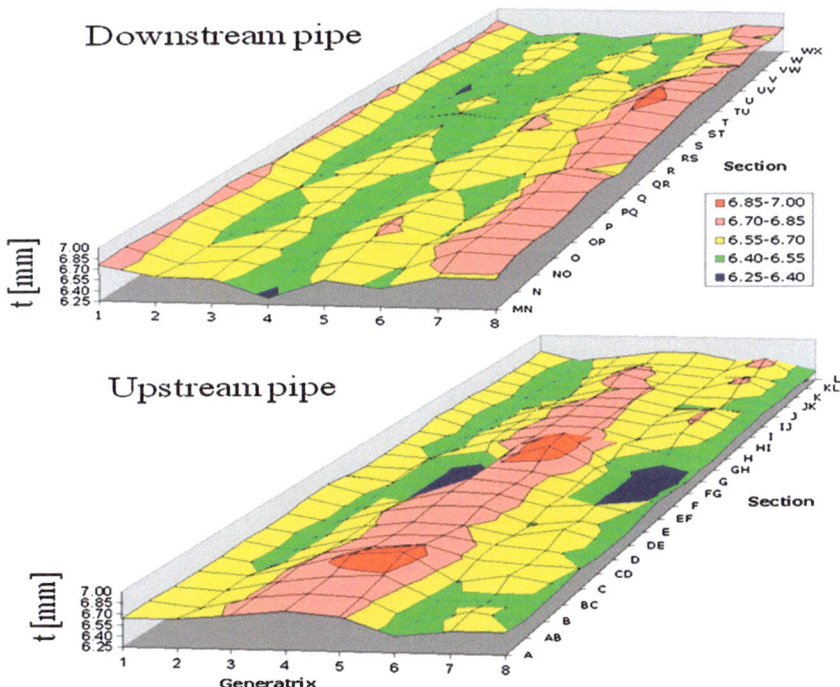

Fig. A.6 Sample #1–5 9/16″ OD 6.55 mm WT AST A-333. Thickness distribution

References

1. Arbocz J, Babcock CD (1969) The effect of general imperfections on the buckling of cylindrical shell. ASME J Appl Mech 36:28–38
2. Arbocz J, Williams JG (1977) Imperfection surveys of a 10-ft diameter shell structure. AIAA J 15:949–956
3. Yeh MK, Kyriakides S (1988) Collapse of deepwater pipelines. ASME J Energy Res Technol 110:1–11
4. Assanelli AP, Toscano RG, Johnson D, Dvorkin EN (2001) Experimental/numerical analysis of the collapse behavior of steel pipes. Eng Comput 17:459–486
5. Press WH, Flannery BP, Teukolsky SA, Veherling WT (1986) Numerical recipes. Cambridge University Press, Cambridge
6. Shunmugam MS (1991) Criteria for computer-aided form evaluation. ASME J Eng Ind 113:233–238
7. Brigham EO (1988) The fast Fourier transform and its applications. Prentice Hall, New Jersey